Simplified Site Engineering
for Architects and Builders

BOOKS BY HARRY PARKER

Simplified Design of Reinforced Concrete
  Second Edition

Simplified Design of Roof Trusses for Architects and Builders
  Second Edition

Simplified Design of Structural Steel
  Second Edition

Simplified Design of Structural Timber

Simplified Engineering for Architects and Builders
  Second Edition

Simplified Mechanics and Strength of Materials

Simplified Site Engineering for Architects and Builders
  By Harry Parker and John W. MacGuire

Kidder-Parker Architects' and Builders' Handbook
  By the late Frank E. Kidder and Harry Parker
  Eighteenth Edition

Materials and Methods of Architectural Construction
  By Harry Parker, the late Charles Merrick Gay, and John W. MacGuire
  Third Edition

NEW YORK · JOHN WILEY & SONS, INC.

London · Chapman & Hall, Limited

Simplified Site Engineering

for

Architects and Builders

BY

HARRY PARKER, M.S.

Emeritus Professor of Architectural Construction
School of Fine Arts
University of Pennsylvania

and

JOHN W. MacGUIRE, B. Arch.

Associate Professor of Achitectural Engineering
School of Fine Arts
University of Pennsylvania

JOHN WILEY & SONS, Inc.

NEW YORK   LONDON

PRINTED IN THE UNITED STATES OF AMERICA

PREFACE

In the preparation of a set of architectural drawings for a building project, numerous problems on many subjects must be solved. Many of these are routine procedure and can be solved by any architect. There is, however, a certain type of problem that requires the attention of someone who has specific knowledge of problems that arise in the analysis of building sites and the preparation of the site plan. Many architectural offices lack men who are qualified for this work, and this book has been written to explain and solve problems of this type. Those who have the knowledge afforded by a course in surveying are equipped to handle many of the problems that are encountered.

This, however, is not just another book on surveying, and no attempt has been made to describe all the many methods of procedure and computations. Simple methods for using surveying instruments are explained; they may be readily applied. Although most architects are seldom required to perform actual surveying, they frequently need the ability to perform the office computations that result from surveying data. This book explains in detail the solution of problems of this nature, and numerous examples are presented. The book will serve also as a review and refresher on subjects that may have been forgotten.

The items contained are both numerous and varied, and the following list enumerates some of the problems that are discussed, illustrated, and solved:

The interpretation of deed descriptions from which site plans are plotted.

The dimensioning of buildings and sites when the angles are other than right angles.

The operation of surveying instruments and the making of surveys and site plans.

The computation of areas of irregular plots.

Dimensioning and laying out of circular curves for driveways and buildings involving arcs of circles.

Vertical curves.

The analysis of contour lines and their manipulation in the solution of grading problems.

The computation of excavation volumes in ground having an uneven surface.

The computation of volumes of cut and fill as indicated by contour lines.

The use of the planimeter.

Maximum and minimum grades for driveways, sidewalks, play areas, etc.

The computation of drainage pipe sizes.

Staking out buildings and driveways.

Items to be considered in the selection of a site.

A check list for site plans.

In the solution of mathematical expressions one can, if he so desires, multiply and divide by the methods used in everyday arithmetic. But, in the dimensioning of plans and sites, the use of logarithms affords both a high degree of accuracy and a great saving of time. To familiarize the user of this book with this valuable tool, the basic principles of logarithms are explained and a five-place log table of numbers is provided. Trigonometric functions are included, with explanations of their uses when required in the dimensioning of drawings. One need have no knowledge of advanced mathematics to understand the computations included in this book; arithmetic and high-school algebra afford adequate preparation. A unique feature of this book is that problems requiring computations are accompanied with logarithmic computations shown in detail.

Following the format of the other books in the "Simplified Series," this book contains concise explanations of procedures illustrated by the solution of practical problems. Included also are problems to be solved by the student. Because of the manner of presentation and the arrangement of material, this book is appropriate for use in classrooms as well as for home study.

Thanks are extended to the Keuffel and Esser Company for permission to adapt from their catalogues the cuts of instruments shown in Figs. 4-1, 4-2, and 13-5.

The authors are well aware that many of the problems relating to site engineering demand the services of a qualified engineer. There are, however, many problems of this nature that arise constantly that may be readily solved by the architect or builder. The purpose of this book is to provide the knowledge for such problems.

HARRY PARKER

JOHN W. MACGUIRE

Philadelphia, Pa.
July, 1954

SUGGESTIONS

The first two chapters of this book explain the fundamental principles of logarithms and trigonometry. If you have a working knowledge of these subjects, you are fortunate; you can begin with Chapter 3, "Measuring Distances."

It is surprising, however, that many men have never studied logarithms or, having once mastered the principles, have forgotten their application in the mathematical processes of multiplication and division. Given ample time and patience, anyone can perform long series of multiplications and divisions by using the everyday procedures learned in arithmetic. But solving expressions with several terms in the numerator and denominator by these methods may be a long and arduous task. However, with the aid of a system of numbers called logarithms, the computations are greatly simplified and much time is saved. For example, extracting the root of a number by arithmetic may be an irksome undertaking but with the use of logarithms the procedure is surprisingly simple. Consequently, it is earnestly recommended that the various steps explained in Chapter 1 be mastered. This cannot be done quickly, but the knowledge acquired will be found to be most useful. The illustrative examples in the book employ logarithms and, in each instance, the procedure is explained in detail.

For those who have little or no knowledge of the subjects presented it is necessary that the material in each chapter be thoroughly understood before continuing with that which follows. The sequence in which the various items are presented is important, and each subject involves material explained in the preceding chapters.

The dimensions on drawings must be accurate. As an aid in preventing errors, form the habit of checking the computations in each step, for an error in one step results in errors in all the succeeding computations.

Do not be hasty or impatient; be thorough and painstaking. You will be rewarded with valuable working knowledge.

CONTENTS

CHAPTER 1

LOGARITHMS

CHAPTER 2

TRIGONOMETRY

CHAPTER 5

LAND SURVEYS

CHAPTER 6

SURVEY COMPUTATIONS

CHAPTER 7

COMPUTATION OF AREAS

CHAPTER 8

MISCELLANEOUS SURVEYING PROBLEMS

CHAPTER 9

CIRCULAR CURVES

CHAPTER 10

LEVELLING

CHAPTER 11

CONTOURS

CHAPTER 12

USES OF CONTOURS

CHAPTER 13

COMPUTATIONS FOR CUT AND FILL

CHAPTER 14

VERTICAL CURVES

CHAPTER 15

DRAINAGE AND GRADING

CONTENTS

CHAPTER 16

STAKING OUT BUILDINGS AND ROADS

CHAPTER 17

SITE SELECTION CONSIDERATIONS

CHAPTER 18

CHECK LIST FOR SITE PLANS

CHAPTER 1

LOGARITHMS

1-1. The Use of Logarithms. In the computations relating to surveying and site engineering, the use of logarithms permits a high degree of accuracy and is a great time saver when compared with the common methods used in arithmetic. The use of numbers called logarithms permits us to change the processes of multiplication and division to the shorter operations of addition and subtraction.

To those who are unfamiliar with logarithms, the study may, at first, seem tedious. Do not be discouraged. The items in this book are arranged in a special sequence. Master each article before proceeding with that which follows. The only method of acquiring facility in using logarithms is by the solution of problem after problem. This cannot be done quickly but, if you are persistent, you will be well repaid for your efforts. The use of logarithms is an invaluable tool in performing computations.

1-2. The Two Fundamental Operations in Mathematics. All mathematical operations may be reduced to addition and subtraction. Suppose, for example, that we have no knowledge of the multiplication tables and that we wish to multiply 5 by 4. To do this we may set down the number 5 four times, and add. Thus, $5 + 5 + 5 + 5 = 20$, which is the result of multiplying 5 by 4. To divide 19 by 6, we can set down 19 and successively subtract 6. Thus, $19 - 6 = 13$, $13 - 6 = 7$, $7 - 6 = 1$. Hence, we see that 6 has been subtracted from 19 three times, and that there is a remainder of 1. Consequently, 19 divided by 6 is 3 with 1 remaining; $19 \div 6 = 3\frac{1}{6}$.

1-3. Powers of a Number. Suppose we wish to raise a number to the third power. This means that we multiply the number by itself three times. The number 5 raised to the third power is written 5^3, and is equal to $5 \times 5 \times 5$, or 125. The small 3 written to the right and above the number is called the *exponent* of the number. A number raised to the first power is the number itself. Thus, $5^1 = 5$.

1-4. Powers of the Number 10. We know that $10^2 = 100$. In this equation, if we say that 10 is the *base*, 100 is the *number*, then 2, the *exponent*, is the *logarithm* of the number. In using logarithms we will be dealing constantly with powers of the number 10. Logarithms to the base 10 are called *common logarithms*. For various powers of 10.

$$10^1 = 10$$

$$10^2 = 100$$

$$10^3 = 1,000$$

$$10^4 = 10,000, \text{ etc.}$$

Notice that 10^3 is equal to 1 followed by 3 zeros. Also, 10^7 would be equal to 1 followed by 7 zeros, or 10,000,000.

1-5. Logarithms. With 10 as the base, *a logarithm of a number is the power to which* 10 *must be raised to give that number*. From Art. 1-4 we see that the logarithm of 1,000 is 3 because 10 must be raised to the third power to give 1,000. We say, "the logarithm of 1,000 is 3"; it is written "log 1,000 = 3." Thus

$$\log 10 \quad\;\; = 1$$

$$\log 100 \quad\; = 2$$

$$\log 1,000 \;\; = 3$$

$$\log 10,000 = 4, \text{ etc.}$$

The logarithms of these numbers are apparent, but most of the numbers with which we deal are not so obvious. Consider, for example, the number 423.

The logarithm of the number 423 is the power to which 10 must be raised to give 423. Now, since 423 lies between 100 and 1,000, the logarithm of 423 will lie between 2 and 3. In other words, the logarithm of 423 will be 2 plus some decimal. As will be explained later, this decimal is 0.62634. Therefore, we say

$$\log 423 = 2.62634$$

Note that a logarithm is composed of *two parts*, that to the *left* of the decimal point and that to the *right*.

1-6. **The Characteristic.** The portion of the logarithm that falls to the *left* of the decimal point is called the *characteristic*. *The characteristic is always one less than the number of digits in the given number.* Thus, since there are 3 digits in the number 423, the characteristic of 423 is 3 minus 1, or 2.

Note particularly that the *number of digits* in a number refers to the number of places *to the left of the decimal point.* Thus

423 contains 3 digits
42,300 contains 5 digits
but
423.36 contains 3 digits
42.3 contains 2 digits

1-7. **The Mantissa.** That portion of a logarithm that lies to the *right* of the decimal point is called the *mantissa*. This part of a logarithm is found by referring to a table of logarithms, such as Table 1 in the Appendix, a five-place table of logarithms of numbers.

1-8. **Logarithms of Numbers Greater Than 10.** To determine the logarithm of a number two steps are required.

STEP 1. Determine the characteristic of the number by subtracting 1 from the number of digits (to the left of the decimal point) in the number.

STEP 2. Find the mantissa by referring to a table of logarithms of numbers, such as Table 1 in the Appendix.

EXAMPLES

EXAMPLE. Find the logarithm of 722.3.

SOLUTION. STEP 1. The number 722.3 contains three digits and, therefore, the characteristic is 3 minus 1, or 2.

STEP 2. In Table 1, page 208, find the number 722 in the column headed N (the number). From this figure move to the right under the column headed 3. Here we find 5872. The four figures in the 3 column are the *last* four figures of the five-place decimal part of the power of 10, and the first figure of the decimal is 8, shown on the 720 line. Table 1 shows that the *first* figure of the mantissa of all numbers between 7200 and 7299 is 8 but, to avoid repetition, it is purposely omitted in the tabulation. Hence the mantissa of the logarithm 722.3 is 85872. In Step 1 the characteristic was found to be 2; therefore,

$$\log 722.3 = 2.85872$$

This is another way of saying $722.3 = 10^{2.85872}$.

EXAMPLE. Find the logarithm of 42.3.

SOLUTION. STEP 1. Since the number contains two digits, the characteristic is 2 minus 1, or 1.

STEP 2. Refer to Table 1, and on page 202 find 423 in the N column. The number 42.3 may be written 42.30; therefore, in the 0 column we find the mantissa of the logarithm of 42.3 to be 62634. Therefore,

$$\log 42.3 = 1.62634$$

It is important to realize that the mantissa of a logarithm is no indication of the magnitude of the number. *The magnitude of the number is indicated by the characteristic.* Thus

$$\log 42.3 \quad = 1.62634$$

$$\log 423 \quad = 2.62634$$

$$\log 42,300 = 4.62634$$

EXAMPLE. Find the logarithm of 316.3.

SOLUTION. STEP 1. There are three digits to the left of the decimal point and, therefore, the characteristic of the logarithm is 2.

STEP 2. On the 316 line of Table 1, page 200, under the 3 column we find the mantissa to be 50010. Therefore, $\log 316.3 = 2.50010$. Note that the first figure of the decimal is 5, whereas $\log 316.2 = 2.49996$. The first figure of the decimal changes from 4 to 5 between the numbers 316.2 and 316.3.

Note that $\log 68.43 = 1.83525$. In the log table the fifth place of the mantissa has the 5 underscored. This indicates that the fifth place is slightly less than 5. A seven-place log table shows that $\log 68.43 = 1.8352465$.

Verify the following logarithms:

$$\log 71.74 \ = 1.85576$$
$$\log 496.3 \ = 2.69574$$
$$\log 15.45 \ = 1.18893$$
$$\log 1626 \ \ = 3.21112$$
$$\log 54,620 = 4.73735$$
$$\log 1007 \ \ = 3.00303$$

PROBLEMS

Find the logarithms of the following numbers by use of Table 1:

1-8-A. 763.4.   **1-8-C.** 4,765.   **1-8-E.** 45,120.   **1-8-G.** 85.95.   **1-8-I.** 500.
1-8-B. 23.92.   **1-8-D.** 300.5.   **1-8-F.** 19.95.   **1-8-H.** 50,120.   **1-8-J.** 1,001.

1-9. Logarithms of Numbers between 0 and 10. The value of any number with an exponent of 0 is 1. Hence, $10^0 = 1$. Consequently, $\log 1 = 0$, and the logarithm of any number between 1 and 10 will have a characteristic of 0. The rule given in Art. 1-6 states that the characteristic is always one less than the number of digits to the left of the decimal point in the given number. Therefore, if the number of digits is 1, 1 less than 1 is 0, and the characteristic of the logarithm will be 0.

EXAMPLE

EXAMPLE.   Find the logarithm of 9.83.

SOLUTION.   STEP 1.   Applying the above rule, 1 minus $1 = 0$, therefore 0 will be the characteristic.

STEP 2.   On referring to Table 1, page 213, we find the mantissa of 983 to be 99255. Therefore, $\log 9.83 = 0.99255$.

1-10. Logarithms of Numbers Less Than 1. If a number is less than 1, the characteristic of the logarithm of the number is a negative integer. *When no digits occur to the left of the decimal point (the number being a decimal), count the number of zeros to the right of the decimal point and add* 1. *This will give the minus characteristic of the number.*

Suppose we wish to find the logarithm of 0.0423. In accordance with the above rule, the number of zeros to the right of the decimal point is 1. Then $1 + 1 = 2$, the *minus* characteristic, -2. The mantissa of 423 is found in Table 1 to be 62634. Therefore,

$$\log 0.0423 = -2.62634$$

Note particularly that *only the characteristic is negative, not the mantissa.*

For convenience in computation, *if the characteristic is negative, it is changed to its equivalent value.* The procedure is as follows:

The logarithm of a number can be written as the sum of an integer plus a decimal fraction:

$$\log \text{ of a number} = \text{characteristic} + \text{mantissa}$$

$$\log 563.1 \qquad = 2.75059 = 2 + 0.75059$$

$$\log 0.0423 = -2 + .62634 = (8 - 10) + .62634 = 8.62634 - 10$$

That is, $\log 0.0423 = -2.62634$ which, written in its equivalent form, is $8.62634 - 10$. *The minus sign applies only to the characteristic, not the mantissa.*

The following logarithms illustrate the application of the rules given in Art. 1-8, 1-9, and 1-10.

$$\log 423 \qquad = \quad 2.62634$$

$$\log 42.3 \qquad = \quad 1.62634$$

$$\log 4.23 \qquad = \quad 0.62634$$

$$\log 0.423 \qquad = -1.62634 \quad \text{or} \quad 9.62634 - 10$$

$$\log 0.0423 \qquad = -2.62634 \quad \text{or} \quad 8.62634 - 10$$

$$\log 0.00423 \qquad = -3.62634 \quad \text{or} \quad 7.62634 - 10$$

$$\log 0.00000423 = -6.62634 \quad \text{or} \quad 4.62634 - 10$$

Note that a logarithm may be either positive, negative, or zero. Thus, if N is the number,

$N =$	.0001	.001	.01	.1	1	10	100	1,000	10,000	100,000
$\log N =$	-4	-3	-2	-1	0	1	2	3	4	5

The following examples illustrate the foregoing principles. Verify these values by the use of Table 1.

$$\log 0.2315 \ = \ -1.36455 \quad \text{or} \quad 9.36455 - 10$$
$$\log 0.0576 \ = \ -2.76042 \quad \text{or} \quad 8.76042 - 10$$
$$\log 0.1005 \ = \ -1.00217 \quad \text{or} \quad 9.00217 - 10$$
$$\log 100.2 \quad = \quad \ \ \ 2.00087$$
$$\log 1.007 \quad = \quad \ \ \ 0.00303$$
$$\log 0.00705 = \ -3.84819 \quad \text{or} \quad 7.84819 - 10$$

PROBLEMS

By the use of Table 1, find the logarithms of the following numbers:

1-10-A. 340.2.	**1-10-E.** 0.8964.	**1-10-I.** 5,000.	**1-10-M.** 6.924.
1-10-B. 1.367.	**1-10-F.** 0.0421.	**1-10-J.** 98,200.	**1-10-N.** 0.00673.
1-10-C. 0.4162.	**1-10-G.** 567.6.	**1-10-K.** 467.5.	**1-10-O.** 1.291.
1-10-D. 23.68.	**1-10-H.** 0.00032.	**1-10-L.** 0.9261.	**1-10-P.** 1.004.

1-11. Interpolation. Table 1, a five-place log table, gives the mantissas of the logarithms of numbers correct to four decimal places provided the number has no more than four significant digits. For the problems given in this book a five-place table of logarithms gives sufficient accuracy but, where a greater number of significant figures is required in the answer or is given in the data, a seven-place table should be employed. Except in special instances, the mantissas given in log tables are endless decimal fractions and consequently the last figure is not exact. The logarithms of numbers with more than four significant digits may be found by interpolation, but an error, although slight, is unavoidable. Interpolation is seldom required for the problems in this book, but the following examples explain the procedure.

EXAMPLES

EXAMPLE. Find the logarithm of 454.35.
SOLUTION. By the use of Table 1 we find

$$\log 454.30 = 2.65734$$
$$\log 454.35 = \quad \ ?$$
$$\log 454.40 = 2.65744$$

It is seen that log 454.35 will be between 2.65734 and 2.65744.
Since 0.35 is $\frac{5}{10}$ of the way from 0.30 to 0.40, log 454.35 may,
with sufficient accuracy, be assumed to be $\frac{5}{10}$ of the way from
log 454.30 to log 454.40.   The tabular difference between 0.65734
and 0.65744 is 0.00010, and $\frac{5}{10}$ of 0.00010 is 0.5×0.00010, or
0.00005.   Therefore, log 454.35 = 2.65734 + 0.00005 or 2.65739.

EXAMPLE.   Find the logarithm of 87.476.
SOLUTION.   We find, by use of Table 1,

$$\log 87.470 = 1.94186$$

$$\log 87.476 = \qquad ?$$

$$\log 87.480 = 1.94191$$

Since 76 is $\frac{6}{10}$ of the way from 70 to 80, log 87.476 is $\frac{6}{10}$ of the
way from log 87.470 to log 87.480.

The tabular difference between 1.94186 and 1.94191 is 0.00005.
$\frac{6}{10}$ of 0.00005 is 0.00003; consequently

$$\log 87.476 = 1.94186 + 0.00003 \quad \text{or} \quad 1.94189$$

PROBLEMS

Find, by interpolation, the logarithms of the following numbers:

1-11-A. 13,424.   **1-11-D.** 46.628.   **1-11-G.** 2,400,400.   **1-11-J.** 12,374.
1-11-B. 5.6789.   **1-11-E.** 0.75426.   **1-11-H.** 0.0019862.   **1-11-K.** 9.4005.
1-11-C. 3967.1.   **1-11-F.** 0.021678.   **1-11-I.** 6,000,200.   **1-11-L.** 721.65.

1-12. Finding the Number That Corresponds to a Given Logarithm.   The number which corresponds to a certain logarithm is
called the *antilogarithm*, or antilog, of the logarithm.   If log 585.3
= 2.76738, 585.3 is the antilogarithm of 2.76738.   The process of
finding an antilog is the reverse of that of finding a logarithm.
Remember that the *mantissa* gives the sequence of the numbers
and the *characteristic* indicates the magnitude; each must be found
separately.   To find the antilogarithm of a number two steps are
necessary.

STEP 1.   Consider first *only the mantissa* of the logarithm.   Refer
to the table of logarithms, find, and set down the number corre-

FINDING THE ANTILOGARITHM 9

sponding to this mantissa, disregarding the position of the decimal point.

STEP 2. Now observe the characteristic of the logarithm. This enables us to determine the position of the decimal point in the number and we employ, in reverse, the rules given for finding the logarithm of a number.

EXAMPLES

EXAMPLE. Find the number whose logarithm is 2.53807. This is another way of saying "find the antilog of 2.53807."

SOLUTION. STEP 1. In Table 1, page 200, we find that the number 3452 corresponds to the mantissa .53807.

STEP 2. The characteristic of the logarithm 2.53807 is 2. Then, in accordance with Art. 1-8, we add 1 to the characteristic to find the number of digits to the left of the decimal point. Thus $2 + 1 = 3$. This enables us to determine the position of the decimal point in the sequence of figures 3452. Hence the antilog of the logarithm 2.53807 is 345.2, or log 345.2 = 2.53807.

EXAMPLE. Find the number corresponding to the logarithm 1.91765.

SOLUTION. STEP 1. On page 210 of Table 1 we do not find the mantissa 91765. The two closest figures are 91761 and 91766. We will take 91766 since it is closer than 91761. The number corresponding to the mantissa 91766 is 8273.

STEP 2. The characteristic of the logarithm 1.91765 is 1. Hence, $1 + 1 = 2$, the number of digits to the left of the decimal point in the antilog. Therefore, the antilog is 82.73. If we interpolate, as explained in Art. 1-11, the antilog is 82.728, a more accurate figure.

To illustrate the above explanation, verify the following:

The antilog of the logarithm 1.34104   = 21.93

The antilog of the logarithm 3.65919   = 4562

The antilog of the logarithm 0.27911   = 1.902

The antilog of the logarithm $9.62941 - 10 = 0.4260$

The antilog of the logarithm $7.74817 - 10 = 0.005600$

PROBLEMS

Find the antilogs of the following logarithms:

1-12-A. 2.76057.	**1-12-E.** 1.07954.
1-12-B. 3.94002.	**1-12-F.** 7.83404 -10.
1-12-C. 0.20978.	**1-12-G.** 0.00346.
1-12-D. 9.71029 -10.	**1-12-H.** 8.07225 -10.

1-13. Multiplication by the Use of Logarithms. Suppose we wish to multiply 100 by 1,000. We may say $100 \times 1,000 = 10^2 \times 10^3 = 10^5$ which, of course, is 100,000. The logarithms of 100 and 1,000 are 2 and 3, respectively. Therefore, *to multiply one number by another we add the logarithms of the numbers, and then find the antilog of their sum.*

EXAMPLES

EXAMPLE. Compute the value of A when $A = 423 \times 9.83$.
SOLUTION. From Table 1,

$$\log 423 = 2.62634$$

$$\log 9.83 = 0.99255$$

Adding, $\qquad\qquad \log A = 3.61889$

The antilog of logarithm 3.61889 is 4158, the value of A. The above computation is actually

$$10^{2.62634} + 10^{0.99255} = 10^{(2.62634+0.99255)} = 10^{3.61889}$$

Note that we have found the result of multiplying two numbers by adding their logarithms.

EXAMPLE. Find the product of $23.1 \times 0.7854 \times 5001$.
SOLUTION. From Table 1,

$$\log 23.1 = 1.36361$$

$$\log 0.7854 = 9.89509 - 10$$

$$\log 5001 = 3.69906$$

Adding, $\qquad\qquad 14.95776 - 10 = 4.95776$

The antilog of 4.95776 is 90,730, the product of 23.1 × 0.7854 × 5001.   Note that the saving in time, in computations employing logarithms, increases with the number of factors involved.

Attention is called to log 0.7854; it is, of course, −1.89509. When we add a logarithm having a negative characteristic, we write it in an equivalent form.   In this instance we add 10 to the characteristic (−1 + 10 = 9), and then subtract 10 to compensate for the 10 that was added.

Note also that the sum of the logarithms in the last example is 14.95776 −10.   If we add the −10 to the characteristic, 14, we have 4, the characteristic of the logarithm.

In order to gain facility in using the log tables, it is good practice to verify the logarithms and antilogs that are given in the examples in this textbook.   It is important, also, to form the habit of presenting the computations in a neat and orderly fashion. The following illustration shows a method of procedure which has been found to be helpful.

EXAMPLE

EXAMPLE.   Let it be required to find the product of 5.3 × 0.00236 × 53710.

SOLUTION.   STEP 1.   First set down all the factors and their *characteristics*.

$$\log 5.3 \quad\quad = 0$$
$$\log 0.00236 = 7 - 10$$
$$\log 53710 \quad = 4$$

STEP 2.   Refer to the log tables and fill in the *mantissas*.

$$\log 5.3 \quad\quad = 0.72428$$
$$\log 0.00236 = 7.37291 - 10$$
$$\log 53710 \quad = 4.73006$$

STEP 3.   Perform the addition.

$$\log 5.3 \quad\quad = \quad 0.72428$$
$$\log 0.00236 = \quad 7.37291 - 10$$
$$\log 53710 \quad = \quad 4.73006$$

$$\overline{\quad\quad\quad\quad\quad\quad\quad\quad\quad}$$

$$12.82725 - 10 = 2.82725$$

STEP 4. Find the antilog of the sum of the logs found in Step 3.

Antilog = 671.8   *Ans.*

These four steps show the sequence of the successive stages in the computations and, of course, are not to be recopied in each step. If several computations are to be made in the solution of a problem it will be helpful to complete Step 1 for all the computations, then Step 2 for all the computations, and so on. In this manner all the work of finding the logs will be done at one time and all the antilogs will be found at one time. This will avoid much lost motion in referring back and forth between log tables and the computation sheet.

By the use of logarithms, verify the following:

$$259.1 \times 72.3 \qquad\qquad = 18{,}730$$

$$0.9983 \times 276.8 \qquad\qquad = 276.3$$

$$823{,}000 \times 54.31 \qquad\qquad = 44{,}700{,}000$$

$$781 \times 0.00346 \qquad\qquad = 2.702$$

$$0.1782 \times 0.0006215 \qquad = 0.0001108$$

$$4{,}231 \times 0.2683 \times 0.0764 \qquad = 86.73$$

$$28.25 \times 12 \times 0.0000067 \qquad = 0.002271$$

$$256 \times 0.07315 \times 5{,}739 \times 26.12 = 2{,}807{,}000$$

PROBLEMS

By the use of logarithms, solve for x in the following problems:

1-13-A. $x = 29.95 \times 679.1.$
1-13-B. $x = 5.429 \times 0.0234.$
1-13-C. $x = 5{,}670 \times 921 \times 0.056.$
1-13-D. $x = 0.9126 \times 562.5 \times 0.0123.$
1-13-E. $x = 90.67 \times 0.0018 \times 70.42 \times 0.023.$
1-13-F. $x = 1{,}111 \times 0.0674 \times 28.84 \times 914.6.$

1-14. **Division by the Use of Logarithms.** As multiplication is an extended addition, division is an extended subtraction. 100,000 divided by 100 may be written $10^5 \div 10^2$, which we know to be 10^3. We may write $10^5 \div 10^2 = 10^{(5-2)} = 10^3$, or 1,000. Since, to divide a number (the dividend) by another number (the divisor), we subtract the exponents, we may say: *to divide one number by another number subtract the logarithm of the divisor from the logarithm of the dividend; their difference will be the logarithm of the quotient.*

EXAMPLE

EXAMPLE. Divide, by the use of logarithms, 384.2 by 26.1.
SOLUTION

$$\log 384.2 = 2.58456$$
$$\log 26.1 = 1.41664$$

Subtracting, $= 1.16792$, antilog $= 14.72$ *Ans.*

If we wish to subtract a larger logarithm from a smaller one, add 10 to the characteristic of the smaller logarithm and then subtract 10 to compensate for the 10 that was added.

EXAMPLE

EXAMPLE. Divide, by the use of logarithms, 492 by 8434.
SOLUTION

$$\log 492 = 2.69197 = 12.69197 - 10$$
$$\log 8434 = 3.92603$$

Subtracting, $= 8.76594 - 10 = -2.76594$

Antilog $= 0.05834$ *Ans.*

The last two examples show how division of two numbers may be performed by subtracting their logarithms. *A more convenient method of performing division is to make use of cologarithms.*

1-15. **Cologarithms.** If a given number is N, the cologarithm (written colog) of N is the logarithm of $1/N$. The logarithm of $1 = 0$ (Art. 1-9); therefore

$$\text{colog } N = \log \frac{1}{N} = 0 - \log N$$

Therefore, *to find the colog of a number subtract the logarithm of the number from 0.*

EXAMPLE

EXAMPLE. Find the colog of 0.031.
SOLUTION

$$\text{colog } 0.031 = \frac{\log 1}{\log 0.031}$$

$$\log 1 \quad = 0 \quad\quad = 10.00000 - 10$$

$$\log 0.031 \quad = -2.49136 = \quad 8.49136 - 10$$

Subtracting, colog 0.031 $=$ 1.50864

By referring to the log tables, the colog of a number may be set down by mentally subtracting the log of the number from 10.00000 − 10. To do this, begin at the *left* and successively subtract each digit in the logarithm from 9 with the exception of the last digit which is subtracted from 10. Then append −10. This method is applicable only when the characteristic of the logarithm is a positive integer.

EXAMPLE

EXAMPLE. The log of 26.1 is 1.41664; what is the colog of 26.1?
SOLUTION. Beginning at the left and successively subtracting 1, 4, 1, 6, and 6 from 9, we have 8.5833. Then, subtracting 4 from 10, we have 8.58336. Finally, appending −10, we have 8.58336 −10, the colog of 26.1.

The purpose of cologs of numbers is to facilitate computation. To divide by a given number N is the same as multiplying by $1/N$. Therefore, *to divide one number by another, add the cologs of each factor of the denominator to the logs of the factors of the numerator and find the antilog of their sum.*

EXAMPLE

EXAMPLE. Consider the first example in Art. 1-14; compute
$$x = \frac{384.2}{26.1}.$$

SOLUTION

$$x = 384.2 \times \frac{1}{26.1}$$

$$\log 384.2 = \quad 2.58456$$

$\log 26.1 = 1.41664$ hence colog $26.1 = \quad 8.58336 - 10$

Adding, $\qquad \log x = 11.16792 - 10 = 1.16792$

$$x = 14.72 \qquad Ans.$$

Verify the following list of cologs:

$$\text{colog } 56.25 \quad = 8.24988 - 10$$

$$\text{colog } 238.1 \quad = 7.62324 - 10$$

$$\text{colog } 1.235 \quad = 9.90833 - 10$$

$$\text{colog } 7.663 \quad = 9.11560 - 10$$

$$\text{colog } 7,244 \quad = 6.14002 - 10$$

$$\text{colog } 0.178 \quad = 0.74958$$

$$\text{colog } 0.004631 = 2.33433$$

$$\text{colog } 0.05728 \quad = 1.24200$$

PROBLEMS

Find the cologs of the following numbers, writing them directly from the table of logs by subtracting mentally:

1-15-A. 417.6.   **1-15-D.** 1.421.   **1-15-G.** 1.424.   **1-15-J.** 21.21.
1-15-B. 8,279.   **1-15-E.** 59.66.   **1-15-H.** 0.0063.   **1-15-K.** 1.437.
1-15-C. 21.34.   **1-15-F.** 0.0127.   **1-15-I.** 0.4321.   **1-15-L.** 0.0196.

1-16. **Powers of Numbers.** When we raise a number to a given power we multiply the number by itself a certain number of times. For instance, 5 raised to the third power is written 5^3, signifying that 5 is to be multiplied by itself three times, $5^3 = 5 \times 5 \times 5 = 125$. However, we have found that multiplying a series of numbers is the same as adding their logarithms. Hence $5^3 = 5 \times 5 \times 5 = \log 5 + \log 5 + \log 5$. But this is equivalent to $3 \times \log 5$. In the quantity 5^3, 3 is the exponent. Therefore, *to raise a number to any power, multiply the logarithm of the number by the exponent and find the antilog of the product.*

EXAMPLES

EXAMPLE. Compute $x = 5^3$ by the use of logarithms.
SOLUTION. $\log x = 3 \times \log 5$; $\log 5 = 0.69897$. Hence $3 \times 0.69897 = 2.09691$, the logarithm of x, and $x = 125$, *Ans.*

EXAMPLE. Compute $x = 21.73^2$ by the use of logarithms.
SOLUTION

$$\log 21.73 = 1.33706$$
$$\times 2$$
$$\overline{}$$
$$\log x = 2.67412$$
$$x = 472.2 \qquad Ans.$$

Verify the following equalities by the use of logarithms:

$$17.15^2 = 294.1$$
$$8.23^3 = 557.4$$
$$5.629^5 = 5,651$$
$$0.173^2 = 0.02993$$
$$0.00538^3 = 0.0000001557$$

PROBLEMS

Compute the following by the use of logarithms:

1-16-A. 61^3.	**1-16-D.** 4.21^5.	**1-16-G.** $7,564^2$.	**1-16-J.** 0.096^3.
1-16-B. 927^2.	**1-16-E.** 0.15^2.	**1-16-H.** 9.245^4.	**1-16-K.** 341.9^2.
1-16-C. 1.267^4.	**1-16-F.** 43^3.	**1-16-I.** 7.426^2.	**1-16-L.** 0.634^5.

1-17. Roots of Numbers. The root of a given number is that quantity which, if multiplied by itself a certain number of times, will produce the given number. As an example, the cube root of 125 is written $\sqrt[3]{125}$ and 3 is the indicated root. Since $5 \times 5 \times 5 = 125$, 5 is the cube root of 125. From algebra we know that $\sqrt[3]{125} = 125^{1/3}$. Therefore, since $\frac{1}{3}$ is the exponent of 125, in accordance with Art. 1-16 $\log \sqrt[3]{125} = \frac{1}{3}$ of $\log 125$. Consequently, *to find the root of a number, divide the logarithm of the number by the indicated root and find the antilog of the quotient.*

EXAMPLES

EXAMPLE. Compute, by logarithms, $x = \sqrt[3]{125}$.
SOLUTION

$\log \sqrt[3]{125} = \frac{1}{3} \log 125$

$\log 125 \quad = 2.09691$

$\frac{1}{3} \log 125 = \frac{1}{3} \times 2.09691 = 0.69897 = \log x$

$x \qquad = 5 \quad Ans.$

EXAMPLE. Compute, by logarithms, $x = \sqrt[2]{1,562}$.
SOLUTION

$\log \sqrt[2]{1,562} \quad = \frac{1}{2} \log 1,562$

$\log 1,562 \qquad = 3.19368$

$\frac{1}{2} \times 3.19368 = 1.59684 = \log x$

$x \qquad\qquad = 39.52 \quad Ans.$

EXAMPLE. Compute, by logarithms, $x = \sqrt[5]{0.08412}$.
SOLUTION. $\log \sqrt[5]{0.08412} = \frac{1}{5} \times \log 0.08412.$

In this example the logarithm of 0.08412 will be negative. *Before dividing a negative logarithm by a positive integer, write the logarithm in an equivalent form so that the negative part, when divided by the integer, will give -10 as the quotient.*

$\log 0.08412 = -2.92490 = 8.92490 - 10$

$8.92490 - 10 = 48.92490 - 50 \quad \text{(adding and subtracting 40)}$

$\frac{1}{5} \times 48.92490 - 50 = 9.78498 - 10 = -1.78498 = \log x$

$x = 0.6095 \quad Ans.$

Verify the following equations:

$$\sqrt[2]{25.63} = 5.063$$
$$\sqrt[3]{7,268} = 19.37$$
$$\sqrt[5]{153.6} = 2.737$$
$$\sqrt[2]{0.001732} = 0.04162$$
$$\sqrt[3]{0.09876} = 0.4622$$

PROBLEMS

Compute the following by the use of logarithms:

1-17-A. $\sqrt[2]{421}$. **1-17-C.** $\sqrt[3]{0.857}$. **1-17-E.** $\sqrt[2]{21.23}$. **1-17-G.** $\sqrt[3]{6,453}$.

1-17-B. $\sqrt[3]{43,170}$. **1-17-D.** $\sqrt[4]{0.04163}$. **1-17-F.** $\sqrt[4]{5.64}$. **1-17-H.** $\sqrt[5]{8,568}$.

1-18. Extended Multiplication and Division. One number divided by another number may be expressed as a fraction, the numerator over the denominator. Many engineering formulas result in expressions consisting of a number of terms in both the numerator and denominator. The following equation is an example:

$$x = \frac{5 \times 23,000 \times 10,080,000}{384 \times 29,000,000 \times 183.4}$$

To employ logarithms in the solution of such a problem we may follow the rule given in Art. 1-15: *to divide one number by another number, add the cologs of the factors of the denominator to the logs of the factors of the numerator and find the antilog.*

EXAMPLE

EXAMPLE. Compute the value of x in the above expression by the use of logarithms.

SOLUTION

log 5	=	0.69897
log 23,000	=	4.36173
log 10,080,000	=	7.00346
colog 384	=	7.41567 -10
colog 29,000,000	=	2.53760 -10
colog 183.4	=	7.73660 -10
log x	=	29.75403 -30 = -1.75403
x	=	0.5676 *Ans.*

Other methods are sometimes more appropriate in the solution of problems than the use of logarithms. The above expression, for example, might have been readily solved by using the slide rule. Nevertheless, logarithms are indispensable for most surveying problems and the brief discussion in this chapter is presented to enable the student to solve problems found later in the book.

Verify the following by the use of logarithms:

$$\frac{600,000}{20,000 \times 0.875 \times 22} = 1.558$$

$$\frac{17,000}{120 \times 0.875} = 161.9$$

$$\frac{11,900 \times 10.725}{18,000 \times 16} = 0.4431$$

$$\frac{18,900 \times 9.114 \times 1.02}{20,000 \times 20.91 \times 2} = 0.2101$$

$$\frac{0.4217 \times 0.217}{0.3852 \times 0.956} = 0.2485$$

PROBLEMS

Solve for x, in the following expressions, by the use of logarithms:

1-18-A. $x = \dfrac{834 \times 5,622}{29 \times 8,876}$

1-18-B. $x = \dfrac{0.2 \times 12}{0.5 \times 0.43}$

1-18-C. $x = \dfrac{31,100}{1,500 \times 0.75 \times 6.231}$

1-18-D. $x = \dfrac{3,920 \times 5,674 \times 3,966}{56.43 \times 7,777 \times 423.1}$

1-18-E. $x = \dfrac{39.64 \times 721.9 \times 4,788}{194 \times 3,266}$

1-19. Accuracy. No doubt certain discrepancies have been apparent between the answers found by the use of logarithms and those found by simple multiplication and division. For example, in Art. 1-13 we multiplied 423 by 9.83. The product, using logarithms, was found to be 4,158, whereas, by arithmetic, the exact answer is 4,158.09. The result of logarithmic computations is usually subject to an unavoidable error. This is because the mantissas given in log tables are, except in special cases, endless decimal fractions. Table 1 is a five-place log table. The use of a seven-place table will give greater accuracy in the results. However, in using the five-place table of logarithms to multiply 423 by 9.83, the error is about 0.002% and this is well beyond the limit of accuracy of most of the data given in engineering work.

In certain precise surveys, such as city surveying and triangulation work, it is advisable to use a seven-place log table to obtain the required accuracy. The use of such a table presents no difficulties for the principles involved in its use are exactly the same as those that are presented in this book.

CHAPTER 2

TRIGONOMETRY

2-1. Scope of Work. The material in this chapter explains the basic principles by means of which plane right triangles are solved. A right triangle is a triangle one of whose angles is 90°. A triangle consists of six parts: three sides and three angles. The solution of triangles consists in determining certain sides and angles when other sides and angles are given. If any three parts of a triangle are known, provided that at least one is a side, the remaining three parts may be determined. To solve a right triangle it is necessary that we know, in addition to the right angle, two parts, one of which is a side.

2-2. Graphic Solutions of Triangles. The solution of problems relating to triangles may be performed by drawing to scale the known sides and angles and scaling the unknown parts. The results found by this graphical method are not sufficiently accurate for most purposes, but it is advantageous to keep the method in mind. Frequently such a drawing, when used as a check, reveals large errors that have occurred in the mathematical computations.

2-3. The Right Triangle. Figure 2-1 shows a right triangle with the conventional lettering that identifies the various parts. The

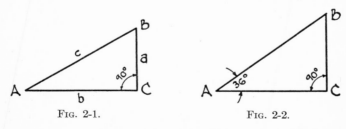

Fig. 2-1.   Fig. 2-2.

three sides are a, b, and c (the hypothenuse). The right angle is angle C, angle A is the interior angle between sides c and b, and

B is the interior angle between sides c and a. The right triangle and the relations of its sides and angles form the basis of trigonometry. Whereas many problems in trigonometry are complex and involved, the ability to solve problems relating to right triangles enables one to perform the computations in connection with surveying that are commonly met with in practice.

2-4. Geometric Principles. Two important principles found in the study of geometry are of great assistance in the solution of triangles.

FIRST. *The sum of the interior angles of a triangle is equal to* 180°. In a right triangle one of the angles is a 90° angle. Consequently, the sum of the remaining two acute angles is 90°. Therefore, if one acute angle is known, this angle subtracted from 90° determines the magnitude of the third angle. As an example, consider the right triangle shown in Fig. 2-2. Angle C is 90° and angle A is 36°. To find angle B, we simply subtract 36° from 90°. Thus angle $B = 90° - 36°$, or 54°, and $A + B + C = 36 + 54 + 90 = 180°$.

SECOND. *In any right triangle the square of the hypothenuse is equal to the sum of the squares of the other two sides.* This is known as the Pythagorean theorem. The hypothenuse is side c, the side

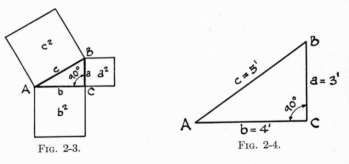

FIG. 2-3.   FIG. 2-4.

opposite the right angle, as shown in Fig. 2-1. It is always the longest side. Referring to Fig. 2-3, $c^2 = a^2 + b^2$. This important principle may be used to determine the unknown side of a right triangle when the remaining two sides are known. Consider, for example, the right triangle shown in Fig. 2-4. Side c, the hypothenuse, is 5′ in length, and side a has a length of 3′. Determine the

length of b, the remaining side.   In accordance with the above
principle,

$$c^2 = a^2 + b^2 \qquad \text{or} \qquad 5^2 = 3^2 + b^2$$

Then $b^2 = 25 - 9 = 16$ or $b = \sqrt{16}$ and $b = 4'$

The right triangle whose sides have 3, 4, and 5 units of length
is sometimes called "the magic triangle."   Without the use of a
surveying instrument, builders frequently lay out right angles with
tapes, using triangles in this proportion: 3, 4, and 5; 15, 20, and
25, etc.

EXAMPLES

EXAMPLE.   Figure 2-5 represents a right triangle in which side
$a = 17.62'$ and side $b = 23.21'$.   Determine the length of side c,

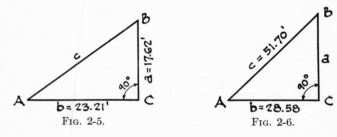

FIG. 2-5.   FIG. 2-6.

the hypothenuse.   With these more complex numbers, the problem may be solved advantageously by the use of logarithms.

SOLUTION

$$c^2 = a^2 + b^2$$

$$c^2 = 23.21^2 + 17.62^2$$

$$\log 23.21 = 1.36568$$
$$\times 2$$
$$\overline{}$$
$$\log 23.21^2 = 2.73136 \qquad \text{antilog} = 538.7$$

$$\log 17.62 = 1.24601$$
$$\times 2$$
$$\overline{}$$
$$\log 17.62^2 = 2.49202 \qquad \text{antilog} = 310.5$$

Then $c^2 = 538.7 + 310.5 = 849.2$

$$c = \sqrt{849.2}$$

$$\log 849.2 = 2.92901$$
$$\div 2$$
$$\overline{}$$
$$1.46451 \qquad \text{antilog} = 29.14$$

Hence $c = 29.14'$ the length of the hypothenuse

EXAMPLE. Figure 2-6 shows a right triangle in which $c = 51.70'$ and $b = 28.58'$. Find the length of side a.

SOLUTION

$$c^2 = a^2 + b^2$$

$$51.70^2 = a^2 + 28.58^2 \qquad \text{hence } a^2 = 51.70^2 - 28.58^2$$

$$\log 51.70 = 1.71349$$
$$\times 2$$
$$\overline{}$$
$$\log 51.70^2 = 3.42698 \qquad \text{antilog} = 2673$$

$$\log 28.58 = 1.45606$$
$$\times 2$$
$$\overline{}$$
$$\log 28.58^2 = 2.91212 \qquad \text{antilog} = 816.8$$

Then $a^2 = 2673 - 816.8 = 1,856.2$, say $1,856$

$$a = \sqrt{1,856}$$

$$\log 1,856 = 3.26858$$
$$\div 2$$
$$\overline{}$$
$$1.63429 \qquad \text{antilog} = 43.08$$

and $a = 43.08'$

For the following right triangles, verify the lengths of the unknown sides.

GIVEN	FIND	ANSWER
$a = 21.36'$ and $b = 60.52'$	c	$c = 64.18'$
$a = 41.23'$ and $b = 13.50'$	c	$c = 43.38'$
$a = 76.10'$ and $c = 82.31'$	b	$b = 31.37'$
$a = 8.36'$ and $c = 96.75'$	b	$b = 96.39'$
$b = 26.28'$ and $c = 35.98'$	a	$a = 24.58'$
$b = 5.23'$ and $c = 5.33'$	a	$a = 1.03'$

PROBLEMS

For the following right triangles the lengths of certain sides are given. Determine, by the use of logarithms, the length of the unknown side.

PROBLEM	GIVEN	FIND
2-4-A.	$a = 61.17'$ and $b = 32.78'$	c
2-4-B.	$a = 4.50'$ and $b = 8.12'$	c
2-4-C.	$a = 52.27'$ and $c = 63.08'$	b
2-4-D.	$a = 14.06'$ and $c = 14.32'$	b
2-4-E.	$b = 10.00'$ and $c = 14.14'$	a
2-4-F.	$b = 59.23'$ and $c = 72.76'$	a

2-5. **Trigonometric Functions of Angles.** Figure 2-7 shows a right triangle in which angle A is 30°. The lengths of sides a, b, and c are 1, $\sqrt{3}$, and 2, respectively. In any right triangle in

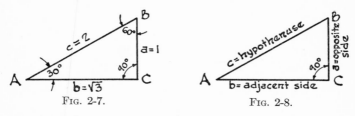

FIG. 2-7.   FIG. 2-8.

which A is 30°, no matter how large or how small the triangle, the ratio of the side opposite angle A to the hypothenuse, a/c, will always be $\frac{1}{2}$; the ratio of the adjacent side to the hypothenuse, b/c, will always be $\sqrt{3}/2$; the ratio of the side opposite angle A to the adjacent side, a/b, will always be $1/\sqrt{3}$, and so on. The ratio of the length of one side to the length of another side is known as a *trigonometric function* of the angle in question. These ratios have specific names, *sine, cosine, tangent,* etc.; but keep in mind that *they are simply ratios.*

In any right triangle there are three sides and two acute angles, an acute angle being an angle between 0° and 90°. Consequently, there are six ratios (functions); they depend on the size of the angle regardless of the size of the triangle. Refer to Fig. 2-8. The names given to the functions are as follows:

$$\text{sine of angle } A \quad = \frac{\text{opposite side}}{\text{hypothenuse}} = \frac{a}{c} \qquad \text{Abbrev.} = \sin A$$

$$\text{cosine of angle } A \quad = \frac{\text{adjacent side}}{\text{hypothenuse}} = \frac{b}{c} \qquad \text{Abbrev.} = \cos A$$

$$\text{tangent of angle } A \quad = \frac{\text{opposite side}}{\text{adjacent side}} = \frac{a}{b} \qquad \text{Abbrev.} = \tan A$$

$$\text{cotangent of angle } A = \frac{\text{adjacent side}}{\text{opposite side}} = \frac{b}{a} \qquad \text{Abbrev.} = \cot A$$

$$\text{secant of angle } A \quad = \frac{\text{hypothenuse}}{\text{adjacent side}} = \frac{c}{b} \qquad \text{Abbrev.} = \sec A$$

$$\text{cosecant of angle } A \quad = \frac{\text{hypothenuse}}{\text{opposite side}} = \frac{c}{a} \qquad \text{Abbrev.} = \csc A$$

The *sine*, *cosine*, and *tangent* are the functions used most frequently, and these ratios, a/c, b/c, and a/b, should be memorized. Note particularly that *the preceding ratios are functions of angle* A.

The other acute angle in the right triangle, Fig. 2-1, is angle B. Now let us consider the functions of angle B. Since, from the above, the sine of an angle is $\dfrac{\text{opposite side}}{\text{hypothenuse}}$, the sine of angle B = b/c. Similarly, the cosine of angle $B = a/c$ and the tangent of angle $B = b/a$. Referring to the conventional system of lettering shown in Fig. 2-1, note the following relationships:

$$\sin A = a/c = \cos B$$
$$\cos A = b/c = \sin B$$
$$\tan A = a/b = \cot B$$
$$\sec A = c/b = \csc B$$
$$\csc A = c/a = \sec B$$
$$\cot A = b/a = \tan B$$

In the right triangle shown in Fig. 2-1, angles A and B are acute angles and angle A + angle B = 90°. Either angle is called the *complement* of the other. For instance, the complement of 40° is 90° − 40°, or 50°. In Fig. 2-1 $\sin A = a/c$ and $\cos B = a/c$; that is, $\sin A = \cos B$. Consider the following pairs of functions: sine and cosine, tangent and cotangent, secant and cosecant. In each pair either function is the *cofunction* of the other one. Any function of angle A equals the cofunction of angle B. Thus $\sin 30° = \cos 60°$, $\tan 26° \, 40' = \cot 63° \, 20'$, and $\sec 50° = \csc 40°$.

Two right triangles are much used by architects. In one the acute angles are each 45° (see Fig. 2-9), and in the other they are

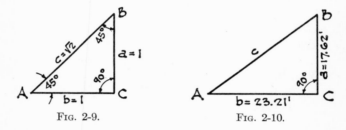

FIG. 2-9.   FIG. 2-10.

30° and 60°, as shown in Fig. 2-7. Side a is one unit of length, and the lengths of the other sides are shown in the figures. From these two figures the accompanying tabulation is readily compiled.

Angle	Sin	Cos	Tan	Cot	Sec	Csc
30°	$\dfrac{1}{2}$	$\dfrac{\sqrt{3}}{2}$	$\dfrac{1}{\sqrt{3}}$	$\sqrt{3}$	$\dfrac{2}{\sqrt{3}}$	2
60°	$\dfrac{\sqrt{3}}{2}$	$\dfrac{1}{2}$	$\sqrt{3}$	$\dfrac{1}{\sqrt{3}}$	2	$\dfrac{2}{\sqrt{3}}$
45°	$\dfrac{1}{\sqrt{2}}$	$\dfrac{1}{\sqrt{2}}$	1	1	$\sqrt{2}$	$\sqrt{2}$

Most of the trigonometric functions shown in the tabulation are fractions. Expressed as decimals, $\frac{1}{2} = 0.5$, $\frac{\sqrt{3}}{2} = 0.86603$, $\frac{1}{\sqrt{3}} = 0.57735$, $\sqrt{3} = 1.73205$, $\frac{2}{\sqrt{3}} = 1.15470$, $\frac{1}{\sqrt{2}} = 0.70711$ and $\sqrt{2} = 1.41421$. Thus, we know that $\sin 30° = 0.5$, $\cos 30° = 0.86603$, $\tan 60° = 1.73205$, $\cos 45° = 0.70711$, etc. These values are called the *natural trigonometric functions*. By referring to a table of natural trigonometric functions we may find directly the function of any angle between 0° and 90°.

2-6. Logarithms of Trigonometric Functions. In the solution of triangles in the following articles the computations involve the multiplication and division of trigonometric functions. Since many of the numbers are complex much time may be saved if the *logarithms of the functions* are used in the computations. Table 2 gives the *logarithms* of the sines and cosines of angles from 0° to 90°, and Table 3 gives the *logarithms* of tangents and cotangents. With these two tables there is no need to find the number corresponding to the natural trigonometric function and then the logarithm of the number. Tables 2 and 3 save a step by giving the logs of the functions directly.

The function of certain angles is less than 1 and therefore the logarithms of these functions have negative characteristics. In order to avoid repetitions in Tables 2 and 3, which give the logarithms of various functions, "-10" has been omitted from all the logarithms, but *we must subtract* 10 *from each logarithm that is given*. As an example, the logarithm of the sine of 30° is given in Table 2 as 9.69897. We know that -10 must be appended, and therefore $\log \sin 30° = 9.69897 - 10$, which we learned in Chapter 1 is -1.69897. Similarly, Table 3 gives $\log \tan 60° = 10.23856$ and, attaching the "-10," $\log \tan 60° = 10.23856 - 10$; this quantity we know to be 0.23856.

Tables 2 and 3 are five-place tables, and the results obtained by their use are accurate up to four places. For problems in which greater accuracy is required, seven-place tables should be used. Usually, five-place log tables give results that are as accurate as the given data.

2-7. Finding Two Angles When Two Sides Are Known. As explained in Art. 2-4, the third side of a right triangle may be found if the lengths of the other two sides are given. Now that we are familiar with the various trigonometric functions explained in Art. 2-5, and have at hand the log tables, we may find the two acute angles when the lengths of two sides are known.

EXAMPLE

EXAMPLE. Figure 2-10 shows a right triangle in which the lengths of two sides are given. Determine the two acute angles A and B.

SOLUTION. Since tan A involves sides a and b, both of which are known, we may write

$$\tan A = \frac{a}{b} = \frac{17.62}{23.21}$$

In Art. 1-15 we found that one number may be divided by another by adding the logs of the factors in the numerator to the cologs of the factors in the denominator. Therefore,

$$\log \tan A = \log 17.62 + \operatorname{colog} 23.21$$

Hence
$$\log 17.62 = 1.24601$$

$$\operatorname{colog} 23.21 = 8.63432 - 10$$

$$\overline{\log \tan A = 9.88033 - 10}$$

Now we refer to Table 3 (page 241) and find the angle corresponding to this logarithm. This is an angle of 37° 12′ (reading to the closest minute). As explained in Art. 2-6, the "−10" must be attached to all the logarithms in this table.

Angle B may be found by again using the tangent.

$$\tan B = \frac{b}{a} = \frac{23.21}{17.62}$$

Hence
$$\log 23.21 = 1.36568$$

$$\operatorname{colog} 17.62 = 8.75399 - 10$$

$$\overline{\log \tan B = 10.11967 - 10}$$

Table 3 shows that the tangents of all angles between 0° and 45° have a characteristic of 9 (minus 10 to be attached). These angles are shown at the tops of the pages. For the tangent of angles greater than 45° we read from the bottom up, the minutes being shown on the *right*-hand side of the table. Hence (page 241), we find an angle of 52° 48′ (reading to the closest minute) for log tan B = 10.11967 −10.

We know that, in a right triangle, angle A + angle B = 90°, and, since 37° 12′ + 52° 48′ = 90° 0°, the above computations are correct.

2-8. Checking Computations. In the solutions of triangles as illustrated in Art. 2-7, there should be no doubt about the correctness of the result for it is always possible to use a different method of solution in order to check the answer. For the problem given in Art. 2-7, angle A was determined first; it was found to be 37° 12′.

Since   angle A + angle B = 90°

                       angle B = 90° − angle A

or   angle B = 90° − 37° 12′

and   angle B = 52° 48′

But this method is no check on the computations involved in solving angle A, for in determining angle B it assumes that angle A is correct. Determining the two angles by different methods and then adding them, knowing that their sum should be 90° 0′, is a valid check on the computations. This procedure was followed in the example. A check should always be made.

For the following right triangles, the lengths of two sides are given and the two acute angles have been determined by computations. Verify the sizes of the angles.

GIVEN	ANSWER
a = 4.72′ and b = 12.26′	angle A = 21° 03′ and angle B = 68° 57′
b = 44.19′ and a = 81.73′	angle A = 61° 36′ and angle B = 28° 24′
a = 7.06′ and c = 11.17′	angle A = 39° 12′ and angle B = 50° 48′
c = 52.25′ and a = 15.92′	angle A = 17° 44′ and angle B = 72° 16′
b = 19.98′ and c = 46.23′	angle A = 64° 24′ and angle B = 25° 36′
c = 54.01′ and b = 38.26′	angle A = 44° 54′ and angle B = 45° 06′

For the above triangles observe that the function of angle A is always the cofunction of angle B; that is, $\sin A = \cos B$ and $\tan A = \cot B$.

PROBLEMS

For the following right triangles the lengths of two sides are given. Compute the sizes of the two acute angles.

Problem	Given	Find
2-8-A.	$a = 27.21'$ and $b = 53.08'$	angle A and angle B
2-8-B.	$b = 4.67'$ and $a = 3.01'$	angle A and angle B
2-8-C.	$a = 10.09'$ and $c = 74.99'$	angle A and angle B
2-8-D.	$c = 4.28'$ and $a = 3.08'$	angle A and angle B
2-8-E.	$b = 61.05'$ and $c = 69.99'$	angle A and angle B
2-8-F.	$c = 35.58'$ and $b = 6.69'$	angle A and angle B

2-9. Solving Right Triangles When One Side and an Acute Angle Are Known. When one side and an acute angle of a right triangle are known, the lengths of the remaining sides may be computed by the use of the trigonometric functions and the logarithmic tables. The other acute angle is found by subtracting the known angle from 90°.

EXAMPLE

EXAMPLE. In Fig. 2-11 angle A is 33° 20′ and side b has a length of 52.33′. Find the remaining parts of the right triangle.

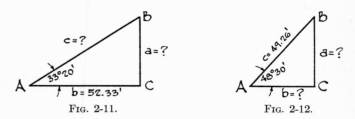

Fig. 2-11.   Fig. 2-12.

SOLUTION. Since the sum of the two acute angles is 90°, 33° 20′ + angle B = 90°, angle B = 90° − 33° 20′, and angle B = 56° 40′.

The tangent of angle A involves sides a and b, Art. 2-5. Since side b is known, we can compute the length of side a.

$$\tan A = \frac{a}{b}$$

Substituting, $\tan 33° 20' = \dfrac{a}{52.33}$ and $a = 52.33 \times \tan 33° 20'$.

We know, from Art. 1-13, that the product of two numbers may be found by adding their logarithms. Therefore,

$\log 52.33 = 1.71875$		Table 1
$\log \tan 33° 20' = 9.81803 - 10$		Table 3

$$\log a = 1.53678$$

antilog $= 34.42'$ the length of side a

Since sides b and c are included in the cosine of angle A, $\cos A$ enables us to compute the length of side c. Thus,

$$\cos A = \frac{b}{c}$$

Substituting, $\cos 33° 20' = \dfrac{52.33}{c}$ and $c = \dfrac{52.33}{\cos 33° 20'}$

In accordance with the principle given in Art. 1-15,

$\log 52.33 = 1.71875$		Table 1
colog $\cos 33° 20' = 0.07806$		Table 2

$$\log c = 1.79681$$

antilog $= 62.63'$ the length of side c

After a was computed we might have determined c by using the sine of angle A. Thus, $\sin A = a/c$. By this formula we could establish c, but it would be based on the assumption that no error had been made in computing a. If a mistake had been made in computing a, c would also have been in error. The method used above in determining c is the proper method for it does not involve the computed side a. *Make it a rule, in determining unknown parts, to use only values given as data.*

By data, we were given the length of b and, by computations, we have determined the lengths of a and c. Now, to check the correctness of these lengths we can apply the principle given in Art. 2-4. Thus,

$$c^2 = a^2 + b^2$$

Substituting, $\qquad 62.63^2 = 34.42^2 + 52.33^2$

$\log 62.63 = 1.79678$	$\log 34.42 = 1.53681$
$\times 2$	$\times 2$
3.59356	3.07362
antilog $= 3{,}922$	antilog $\qquad = 1{,}185$

$$\log 52.33 = 1.71875$$
$$\times 2$$

3.43750

antilog $\qquad = 2{,}738$

Adding, $\qquad\qquad 3{,}923$

Since 3922 and 3923 are in reasonable agreement, the lengths of the computed sides are considered to be sufficiently accurate. The reason for the discrepancy lies in the fact that a five-place log table was used with no recourse to interpolation, as explained in Art. 1-11.

2-10. **Arrangement of Computations.** It is of importance that pains be taken to arrange all computations in a neat, legible, and systematic manner. A habit thus formed will result in a saving of time and will minimize the possibility of error. The practice of checking results will be found to be of great value. This is particularly true in engineering work because the solution of one problem frequently provides the data for ensuing problems. Thus, an error at the beginning, if not detected, may result in wasted effort and confusion. Whenever possible, diagrams should be drawn to scale. Such diagrams frequently reveal the presence of errors.

The following example is presented, without explanatory notes, as a suggested form of computation arrangement.

EXAMPLE

EXAMPLE. In the right triangle shown in Fig. 2-12, angle $A = 48° 30'$ and $c = 49.26'$. Determine angle B and the lengths of sides a and b.

SOLUTION

$$90° 0'$$
$$\text{minus } 48° 30'$$
$$\overline{}$$
$$41° 30' = \text{angle } B$$

$$\sin 48° 30' = \frac{a}{49.26} \qquad a = 49.26 \times \sin 48° 30'$$

$$\log 49.26 = 1.69249$$
$$\log \sin 48° 30' = 9.87446 - 10$$
$$\overline{}$$
$$1.56695 = 36.89' = a$$

$$\cos 48° 30' = \frac{b}{49.26} \qquad b = 49.26 \times \cos 48° 30'$$

$$\log 49.26 = 1.69249$$
$$\log \cos 48° 30' = 9.82126 - 10$$
$$\overline{}$$
$$1.51375 = 32.64' = b$$

Check

$$49.26^2 = 36.89^2 + 32.64^2$$

$$\log 49.26 = 1.69249 \qquad \log 36.89 = 1.56691$$
$$\times 2 \qquad\qquad\qquad \times 2$$
$$\overline{} \qquad\qquad\qquad \overline{}$$
$$3.38498 \qquad\qquad\qquad 3.13382$$
$$= 2426 \qquad\qquad\qquad\qquad\qquad = 1361$$

$$\log 32.64 = 1.51375$$
$$\times 2$$
$$\overline{}$$
$$3.02750$$
$$= 1065$$
$$\overline{}$$
$$2426$$

ANSWER $B = 41° 30'$

  $a = 36.89'$

  $b = 32.64'$

For the following right triangles one acute angle and one side are given. Verify the magnitudes of the remaining parts.

GIVEN	COMPUTED
$A = 30°\ 0'$ and $c = 28.06'$	$B = 60°\ 0'$, $a = 14.03'$, $b = 24.30'$
$B = 22° 30'$ and $c = 73.26'$	$A = 67° 30'$, $a = 67.68'$, $b = 28.04'$
$A = 17° 56'$ and $a = 12.68'$	$B = 72° 04'$, $b = 39.18'$, $c = 41.18'$
$B = 46° 23'$ and $b = 56.73'$	$A = 43° 37'$, $a = 54.05'$, $c = 78.36'$
$A = 72° 41'$ and $b = 8.23'$	$B = 17° 19'$, $a = 26.40'$, $c = 27.65'$
$B = 62° 06'$ and $a = 31.17'$	$A = 27° 54'$, $b = 58.87'$, $c = 66.61'$

PROBLEMS

For each of the following right triangles one acute angle and one side are given. Find the unknown parts.

PROBLEM	GIVEN	FIND
2-10-A.	$A = 45°\ 0'$ and $c = 14.14'$	B, a, and b
2-10-B.	$B = 81° 01'$ and $c = 92.32'$	A, a, and b
2-10-C.	$A = 36° 42'$ and $a = 6.04'$	B, b, and c
2-10-D.	$B = 15° 15'$ and $b = 11.12'$	A, a, and c
2-10-E.	$A = 8° 16'$ and $b = 34.23'$	B, a, and c
2-10-F.	$B = 45° 05'$ and $a = 26.01'$	A, b, and c

2-11. Oblique Triangles. The Sine Law. Whereas the principles involved in the solution of right triangles may be used in solving any triangle, many trigonometric formulas have been

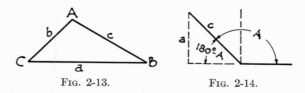

FIG. 2-13.   FIG. 2-14.

derived which simplify the solution of certain problems. Among these formulas is the *sine law*.

THE SINE LAW. *In any triangle the sides are proportional to the sines of the opposite angles.* Figure 2-13 shows any triangle and, in accordance with the sine law,

$$\frac{a}{\sin A} = \frac{b}{\sin B} = \frac{c}{\sin C}$$

In Art. 2-5, referring to Fig. 2-8, a right triangle, we found that

$$\sin A = \frac{\text{opposite side}}{\text{hypothenuse}} = \frac{a}{c}$$

Figure 2-14 shows angle A to be greater than 90°, an *obtuse angle*. From the figure we see that $\sin A = \dfrac{a}{c}$. The numerical value of $\sin A = \sin (180 - A)$.

EXAMPLE

EXAMPLE. In Fig. 2-13, angle $A = 105°$, angle $B = 30°$, and angle $C = 45°$. If side $a = 50'$, find the lengths of b and c.

SOLUTION. Since angle A is an obtuse angle, $\sin A = \sin (180° - 105°)$, or $\sin A = \sin 75°$. Now, applying the sine law,

$$\frac{a}{\sin A} = \frac{b}{\sin B} \qquad \text{or} \qquad b = \frac{50 \sin 30°}{\sin 75°}$$

$$\log 50 = 1.69897$$

$$\log \sin 30° = 9.69897 - 10$$

$$\text{colog} \sin 75° = 0.01506$$

$$\overline{}$$

$$1.41300 = 25.88', \text{ the length of } b$$

Similarly, $\qquad \dfrac{a}{\sin A} = \dfrac{c}{\sin C} \qquad \text{or} \qquad c = \dfrac{50 \sin 45°}{\sin 75°}$

$$\log 50 = 1.69897$$

$$\log \sin 45° = 9.84949 - 10$$

$$\text{colog} \sin 75° = 0.01506$$

$$\overline{}$$

$$1.56352 = 36.60', \text{ the length of } c$$

PROBLEMS

2-11-A. For the triangle shown in Fig. 2-15, find the lengths of sides AB and BC and also angle A.

2-11-B. For the triangle shown in Fig. 2-16, find the lengths of sides AB and BC and also angle B.

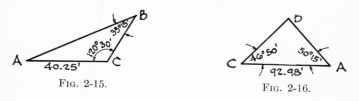

FIG. 2-15.   FIG. 2-16.

2-12. Areas of Triangles. The area of a right triangle may be found by taking ½ the product of the base by the height.

When the lengths of all three sides of a triangle are known, the area of any triangle may be found by use of the following formula:

$$\text{Area} = \sqrt{s(s-a)(s-b)(s-c)}$$

in which a, b, and c are the lengths of the sides of the triangle and $s = \frac{1}{2}(a + b + c)$.

EXAMPLE

EXAMPLE. Compute the area of the triangle shown in Fig. 2-17.

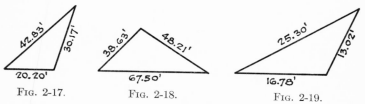

FIG. 2-17.   FIG. 2-18.   FIG. 2-19.

SOLUTION. First, let us compute the magnitude of s.

$$s = \frac{30.17 + 20.20 + 42.83}{2} = \frac{93.20}{2} = 46.60$$

Then,

$$
\begin{array}{ccc}
46.60 & 46.60 & 46.60 \\
-30.17 & -20.20 & -42.83 \\
\hline
(s-a) = \quad 16.43 & (s-b) = \quad 26.40 & (s-c) = \quad 3.77
\end{array}
$$

$$\text{Area} = \sqrt{46.60 \times 16.43 \times 26.40 \times 3.77} = 276.1\text{ft}^2$$

$$\log 46.60 = 1.66839$$

$$\log 16.43 = 1.21564$$

$$\log 26.40 = 1.42160$$

$$\log 3.77 = 0.57634$$

$$\overline{}$$

$$4.88197$$

$$\div 2$$

$$\overline{}$$

$$2.44099 = 276.1$$

PROBLEMS

2-12-A. Compute the area of the triangle shown in Fig. 2-18.

2-12-B. Compute the area of the triangle shown in Fig. 2-19.

CHAPTER 3

MEASURING DISTANCES

3-1. Introduction. It is not expected that the architect or builder will acquire the skill or degree of accuracy of the professional surveyor. Nevertheless, in measuring distances, greater proficiency is obtained when one is familiar with certain basic principles and the common errors that are to be avoided. The tape is used more often by architects and builders than any of the surveyors' instruments and, although other instruments are sometimes used, the measurement of distance is generally accomplished by its use. For measuring distances where the tape cannot be used, trigonometric methods are employed.

3-2. Tapes. Two types of tapes for measuring distances are available. There is the *cloth tape*, a tape made of cloth and in which wires are woven, and the *steel tape*. All tapes are subject to stretching when tension is applied and, because of its greater resistance to deformation, the steel tape is used exclusively for accurate measurements. The steel tape is commonly available

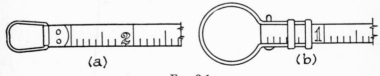

(a)   (b)

Fɪɢ. 3-1.

in 50′ and 100′ lengths and in greater lengths when required. It is a thin ribbon of steel graduated in feet and fractions of a foot. Generally the foot is divided into tenths and hundredths. Before using the tape observe carefully the position of the zero point. Sometimes this point is at the extreme end of the ring at the end of the tape, as shown in Fig. 3-1 (*a*). Sometimes, however, as indicated in Fig. 3-1 (*b*), it is marked on the body of the tape.

The advantage in using the former is that one may hook the end over a pin in the ground or a nail in the wall without the necessity of having someone hold the end of the tape. Much time may be saved and greater accuracy obtained by measuring distances with the aid of an assistant.

3-3. **Chains.** Many old surveys have distances indicated in units of *chains*. A chain consisted of 100 links of heavy steel wire. Each link had a ring at its ends by means of which it was joined to an adjacent link. A chain was 66′ in length and, since there were 100 links in a chain, each link had a length of 0.66′ or 7.92″. Ten square chains equaled 1 acre and, consequently, the use of a chain afforded a convenient measure in surveying land acreage.

By constant use, the wear on the rings of a chain resulted in its lengthening and a subsequent error in dimensions. This is thought to be the reason for the discrepancies between the *District Standards* of length, used in some of our older localities, and the *U. S. Standard* of measurement. The chain is no longer used as a unit of measure in this country. In converting chains, in old deeds or surveys, to feet and inches the following conversion table may be used.

$$7.92 \text{ inches} = 1 \text{ link}$$

$$100 \text{ links} = 1 \text{ chain} = 66 \text{ feet}$$

$$80 \text{ chains} = 1 \text{ mile}$$

3-4. **Care of the Steel Tape.** It is important that certain precautions be taken in the care of the steel tape. Being a thin band of steel, it is brittle and easily broken. It is advisable to wipe the tape at the end of each day with a dry cloth and then give a coating of oil. This precaution avoids rusting which tends to obliterate the markings.

The most important feature of a tape is its length. If, by chance, it is broken it should be repaired only by the manufacturer or by those who specialize in this delicate work. A steel tape properly repaired may be fully as accurate as it was before it was broken.

3-5. **Horizontal Distances.** It is of the utmost importance to bear in mind that *all distances shown on maps and plans are horizontal projections of distances*. The surface of the ground gen-

erally has a slope and seldom lies in a horizontal plane. However,
the distances recorded on surveys or in deed descriptions are
always horizontal distances. Figure 3-2 shows a cross section
taken through the side of a hill. The distance between points A

and B would be measured and
recorded as the distance between
A' and B, actually the horizontal
projection of the distance between
points A and B.

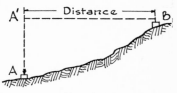

Fig. 3-2.

3-6. **Measuring Horizontal Distances.** Generally, distances are
measured by holding the tape in
a horizontal position and transferring the distance between two
points by means of a plumb line. One man holds the zero end
of the tape over a tack on a stake *at the higher level* while the
other man, holding the tape level with one hand and with a
plumb line in the other, places the plumb bob directly over the
point to be measured and reads the distance at which the plumb
line intersects the tape. This procedure is indicated in Fig. 3-3.

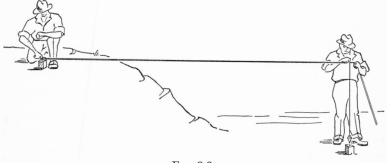

Fig. 3-3.

Experiment will show that *it is much easier to measure downhill
than uphill.*

From the figure, it is obvious that the two ends of the tape
must be held at the same level; the tape must be horizontal if
the measurement is to be accurate. Naturally, there will be a
certain degree of sag in the tape. It is customary to maintain
about a 10-pound tension to measure with a sufficient degree of

accuracy for ordinary work. For precise city surveying a level
is sometimes placed on the tape and a spring balance is employed
to maintain a uniform tension.

Measurements may be made by laying the tape directly on the
slope. Suppose, for example, we wish to determine the horizontal
distance $A'B$ between the two points A and B shown in Fig. 3-4.
The tape is laid on the surface of the slope and the dimension AB
is determined. Then the angle θ is measured by a transit and, by
the principles of trigonometry given in Chapter 2, the distance
$A'B$ is computed. This method of measuring requires consid-

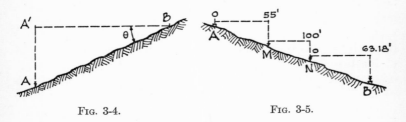

FIG. 3-4.   FIG. 3-5.

erable time and is used only when required by conditions in the
field.

3-7. Measuring on Steep Slopes. For steep slopes it may be
found to be impracticable to hold the tape level for the entire
distance. For such a condition the distance between the two
points to be measured is divided into a number of parts so that
the forward man may hold the tape conveniently at chest
height.

Referring to Fig. 3-5, suppose it is required to measure the
horizontal distance between points A and B, the difference in
the elevation between the two points being approximately 10′.

We begin by having the rear tapeman hold the zero point of
the tape over the upper stake, point A. The head tapeman goes
forward and the rear tapeman directs him to move to the right
or left until he is on the line between points A and B. At a convenient point at some even foot mark, as at M, he places a pin
in the ground and calls this dimension, say 55′, to the rear tapeman. The rear tapeman should repeat this figure aloud to avoid
error. Then the rear tapeman drops the tape, advances to M,
and holds the 55′ mark on the tape over the pin. The forward
tapeman stretches the tape to its full length, say 100′, and places

a pin in the ground at point N. The rear tapeman advances with the tape and holds the zero point over this pin. Then the head tapeman proceeds forward, plumbs over point B, and reads the tape; say the dimension is 63.18′. Thus the horizontal distance between points A and B is 163.18′.

3-8. **Alignment between Points.** When the distance to be measured between two points must be measured in parts, or when it exceeds the length of the tape, it is important that the intervening points be on an approximately straight line. By inspection of Fig. 3-6 it is seen that if the points M and N are not on the line between points A and B (shown in plan) the recorded distance, $AM + MN + NB$, will be greater than the true distance AB. In order to prevent this inaccuracy, the rear tapeman sights from A to B and directs the head tapeman to the

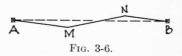

FIG. 3-6.

right or left in establishing the intervening points. In cases in which the termination point is not visible to the rear tapeman, as, for instance, an obstructing knoll, a *ranging pole* is placed in the ground close to the termination point and the rear tapeman sights on this. A ranging pole is usually of wood, octagonal in cross section, having lengths of 6′ to 10′. The pole is enameled red and white in alternate bands and the end is fitted with a steel point.

3-9. **Marking Points on the Survey.** The ends of straight lines on a survey are generally marked by driving wooden stakes into the ground, the exact point being located by a tack or brad driven in the stake. When it is not feasible to drive wooden stakes, metal spikes may be used, the exact point being marked by a center punch in the head of the spike. In rock, concrete pavements, or masonry portions of buildings, a yellow chalk mark is often used, the exact point being located with a black pencil cross on the chalk. For permanent locations, the point may be marked by a small drilled hole.

For temporary locations, and for intermediate points, metal pins, called arrows, are commonly used. These metal pins are approximately $\frac{3}{16}''$ in diameter and 12″ to 15″ in length. They are driven into the ground inclined to the vertical and at right angles to the line between the points being measured. The exact point being marked is the center of the pin at the surface of the ground.

3-10. **Mistakes and Errors.** A *mistake* is the result of faulty operations on the part of the person making the measurement. Acquiring proper work habits will tend to eliminate many mistakes. An *error* is a residual fault in the measuring instrument or in the technique of making the measurement. The use of a tape, the length of which is inaccurate, results in residual errors. Corrections may be made for such errors, and for extremely precise work many different corrections may need to be applied.

3-11. **Cumulative and Compensating Errors.** *Cumulative errors* are constant and uniform errors that affect measurements in the same manner; they consistently either increase or decrease the results of measurements and successively accumulate the error.

Compensating errors, to the contrary, are errors which tend to cancel each other. Whereas cumulative errors are either all plus or all minus, compensating errors in the same series of measurements are both plus and minus. Over a long distance, the ultimate result of cumulative errors in measurements may be considerable but, if the errors are compensating, the resulting error would be comparatively small.

3-12. **Precision.** Absolute accuracy of measurement is an ideal seldom, if ever, attained in reality. The degree of precision required depends on the character of work to be done. As an example, in laying out the wall lines for a building it may be considered sufficiently accurate if the measurements are correct to within $\frac{1}{8}''$. In machining a delicate bearing, however, an error of $1/1,000''$ might be considered to be too great. Obviously, the more precise the measurement, the greater will be the cost of making it. The same care and precision used in surveying a plot of ground in the business center of a large city would be inappropriate for the survey of farm land worth $50.00 per acre.

As related to surveying, precision is the ratio of the error to the distance measured. This ratio is expressed as a fraction. If, for example, a precision of 1/10,000 is permitted, an error of 1′ in a length of 10,000′ would be acceptable. For farm and suburban surveying and for building layouts within the confines of a plot, a precision of 1/5,000 is usually considered to be acceptable. This is an error of approximately $\frac{1}{4}''$ in 100′. City surveys generally require a precision of 1/10,000; sometimes greater precision is required. For this degree of precision the common builders' level is not sufficiently accurate and it becomes necessary to apply certain corrections and refinements which are beyond the scope of this book.

3-13. **Mistakes in Measuring Distances.** Among the common mistakes made in measuring distances is the failure to observe the position of the zero mark on the tape. This is particularly likely to occur if the surveyor changes from one type of tape to another.

In measuring long distances a whole length of tape may be omitted. To avoid this, a pebble or coin is sometimes placed in a pocket for each full tape length recorded. Another aid is to have more than one individual keep the count.

In reading the tape the wrong foot mark may be read, such as 76.92′ instead of 75.92′. Sometimes numbers are read upside down, 6 being mistaken for 9 or 68 instead of 89. Another very common mistake is to transpose figures, such as reading 21.51 instead of 21.15.

It is an excellent habit to estimate mentally the distance to be measured. Such a practice will tend to eliminate many of the larger discrepancies.

3-14. **Errors in Measuring Distances.** Failure to have the tape stretched tight while measuring results in a dimension that is too great. Permitting the tape to bend around bushes, boulders, etc., also results in a faulty measurement. Taking dimensions during a high wind permits the tape to be blown out of line, and this type of work should be avoided when weather conditions are unfavorable. Incorrect alignment of the tape, as explained in Art. 3-8, is another common error, but this particular error may easily be guarded against. The above-mentioned errors all tend to make the recorded measurements too long; they are *cumulative*.

Failure to hold the plumb bob exactly over the point results in a dimension that is either too great or too small. Errors of this type are *compensating*, one error tending to correct a former error.

3-15. **Tapes of Incorrect Length.** One of the most serious sources of error lies in using a tape whose length is false. For example, the markings may indicate the length to be 100′, but, in reality, the true length of the tape may be somewhat more or less than this dimension. If such a tape is used, the resulting measurements will be in error, a *cumulative* error.

Surveyors frequently send one of the tapes in their office to the National Bureau of Standards to have its length checked. On the return of this tape it is kept as a standard by means of which the other tapes are checked so that they may be used in the field.

If the error per tape length is known, a correction figure may be computed. This correction, added or subtracted from a measurement made with the faulty tape, will give the corrected dimension.

RULE 1. *If the tape is* longer *than the standard, the correction must be* added. Suppose that we have a tape marked to be 100′ in length but whose true length is 110′. This, of course, is an exaggerated error and is used for the purpose of illustration;

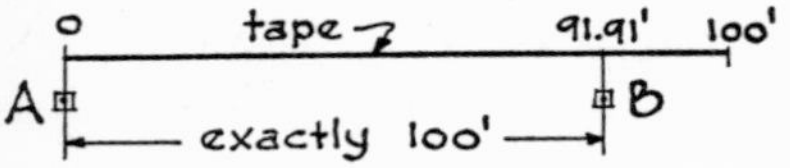

FIG. 3-7. 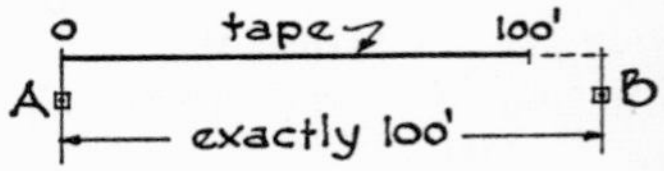   FIG. 3-8.

actually, the errors are usually fractions of an inch. Refer to Fig. 3-7 and note that points A and B are exactly 100′ apart. If we place the zero point of the faulty tape at point A and run out the entire length of the tape, the 100′ mark on the tape will be beyond point B. The reading on the tape at point B would be at approximately 90.91′. Since this tape reading is less than the true dimension, a correction must be *added* to the tape reading to give the exact length.

RULE 2. *If the tape is* shorter *than the standard, the correction must be* subtracted. Let us assume now that we have a tape that, although marked 100′ in length, has a true length of only 90′. In Fig. 3-8 points A and B are exactly 100′ apart. If the zero point on the inaccurate tape is placed at point A and the tape is extended its full length, the 100′ mark on the tape will not reach point B. By use of the faulty tape the measurement from A to B would be 111.11′. Therefore, since this distance is too great, we must *subtract* a correction to the tape reading to find the true measurement.

EXAMPLE

EXAMPLE. The distance between two points when measured with a 100′ steel tape was found to be 323.52′. When compared with a standard, the length of the steel tape was found to be only 99.83' instead of 100'. Compute the true distance between the two points.

SOLUTION

$$100.00' - 99.83' = 0.17'$$

hence,   correction = 0.17 foot per 100 linear feet

or,   correction = 0.0017 foot per linear foot

The correction for the distance 323.52' is 323.52 × 0.0017, or 0.549984', say 0.55'.

Because the tape is *too short*, the correction must be subtracted. Therefore,

$323.52 - 0.55 = 322.97'$ the true distance between the two points

PROBLEMS

The following problems relate to measuring distances with inaccurate tapes. The second column gives the distances measured by the tape, and the third column gives the actual lengths of the tapes when compared with a 100' standard. For each condition compute the true distance.

PROBLEM	MEASURED DISTANCE	ACTUAL LENGTH OF TAPE
3-15-A.	78.63'	100.13'
3-15-B.	153.17'	100.13'
3-15-C.	23.16'	100.09'
3-15-D.	456.73'	99.71'
3-15-E.	83.27'	100.33'
3-15-F.	126.26'	99.94'

3-16. **District Standards.** As noted in Art. 3-3, distances measured by chains often resulted in inaccurate dimensions. This accounts, possibly, for the *District Standards* that exist in certain of the older localities of cities. The District Standards are not in agreement with the U. S. Standard of measurement. When work is done in a locality where District Standards occur, a correction must be added or subtracted to make the dimensions conform to the U. S. Standard. Where such a condition prevails, the degree of correction may be obtained from the engineering department of the municipality. When requesting this information it is necessary to give the exact location of the site, for District Standards may vary in different parts of a community.

District Standards are corrected in the manner described in Art. 3-15.

EXAMPLE

EXAMPLE. For a certain locality in the city of Philadelphia, a district engineer states that the 100.00′ District Standard = 100.25′ U. S. Standard. A distance given on the City Plan is 273.23′. What is the U. S. Standard distance?

SOLUTION. As 100.25 − 100.00 = 0.25′, the correction per 100′, 0.25 ÷ 100 = 0.0025′, the correction per foot. Thus, 273.23 × 0.0025 = 0.683075, say 0.68′, the correction for 273.23′. Because U. S. Standard is greater than District Standard, this correction must be added. Therefore, 273.23 + 0.68 = 273.91′, the true or U. S. Standard dimension.

PROBLEMS

In a certain locality, 100.00′ District Standard equals 100.17′ U. S. Standard. The following distances are in accord with the District Standard. Convert these distances to the U. S. Standard.

3-16-A. 23.13′.	**3-16-C.** 589.48′.
3-16-B. 176.71′.	**3-16-D.** 76.22′.

3-17. **Other Errors in Measurements.** The errors in measurements previously mentioned are those that occur most frequently in the usual surveying work. When a precision of 1/10,000 or more is required, as in more exact city surveying, corrections must be made for other sources of error. Such errors are the result of temperature changes (with the accompanying expansion and contraction of the tape), the application of too much or too little tension on the tape, and the degree of sag. These refinements, however, will generally be beyond the precision required for ordinary surveying work and may safely be ignored.

3-18. **Stations.** In surveying long distances, such as roads or highways, the distances to various points are often indicated as measured continuously from the point of the beginning of the survey. *Stations* are points that are established at each 100′

interval. Thus, a point 723.16′ from the initial point would be noted as "station 7 + 23.16′."

3-19. **Conversion of Decimals of a Foot to Inches.** Because surveyors' instruments are graduated in feet and decimal parts thereof, and because this system of graduations simplifies computations, the surveyor records distances in units of feet and decimal parts of a foot. The architect and builder, however, invariably use a dimensioning system of feet, inches, and fractions of an inch. Because of this, in dimensioning a building or plot plan, it is often necessary to convert decimal parts of a foot to their equivalent in inches and fractions thereof. Sometimes the reverse process is required.

For ordinary surveying, dimensions are given, or required, only to the nearest 1/100 of a foot. For this degree of precision the following system of conversion will be found to be of great convenience. Certain equivalents are known as, for instance,

$$3'' = 0.25' \quad \text{(exact)}$$
$$4'' = 0.33' \quad \text{(approximate)}$$
$$6'' = 0.50' \quad \text{(exact)}$$
$$8'' = 0.67' \quad \text{(approximate)}$$
$$9'' = 0.75' \quad \text{(exact)}$$

Now, since $12'' = 1'$, $1'' = \frac{1}{12}'$, or 0.08333′. Hence, $1'' =$ approximately 0.08′ and, consequently, $\frac{1}{8}'' = 0.01'$, approximately. Using this equality, $\frac{1}{8}'' = 0.01'$, and the above equivalents, we may tabulate inches and their equivalents in decimals of a foot as follows:

$$1'' = 0.08'$$
$$2'' = 0.17' \quad \text{(0.25 less 0.08)}$$
$$3'' = 0.25'$$
$$4'' = 0.33'$$
$$5'' = 0.42' \quad \text{(0.50 less 0.08)}$$
$$6'' = 0.50'$$

$$7'' = 0.58' \qquad (0.50 \text{ plus } 0.08)$$

$$8'' = 0.67'$$

$$9'' = 0.75'$$

$$10'' = 0.83' \qquad (0.75 \text{ plus } 0.08)$$

$$11'' = 0.92' \qquad (1.00 \text{ less } 0.08)$$

$$12'' = 1.00'$$

As for fractions of an inch, we know that $\frac{1}{8}'' = 0.01'$. Then $\frac{1}{4}'' = (2 \times 0.01) = 0.02'$, $\frac{3}{8}'' = (3 \times 0.01) = 0.03'$, etc. Hence, for the decimal equivalent of inches and fractions of an inch, we add or subtract from the nearest whole inch in the above tabulation. For example,

$$3\frac{1}{8}'' = 0.25 + 0.01 = 0.26'$$

$$7\frac{3}{4}'' = 0.67 - 0.02 = 0.65'$$

$$5\frac{5}{8}'' = 0.50 - 0.03 = 0.47'$$

$$11\frac{1}{2}'' = 1.00 - 0.04 = 0.96'$$

Converting decimals of a foot to inches is done in a similar manner; thus,

$$0.53' = 6 + \tfrac{3}{8} \ = 6\tfrac{3}{8}''$$

$$0.23' = 3 - \tfrac{1}{4} \ = 2\tfrac{3}{4}''$$

$$0.68' = 8 + \tfrac{1}{8} \ = 8\tfrac{1}{8}''$$

$$0.89' = 11 - \tfrac{3}{8} \ = 10\tfrac{5}{8}''$$

With practice these conversions may be made mentally and tables may be dispensed with.

PROBLEMS

3-19-A. Convert the following linear dimensions to feet and decimals of a foot: 129' 2½"; 75' 0⅝"; 23' 9¾"; 351' 7⅝"; 17' 4⅜"; 183' 2⅛".

3-19-B. Convert the following dimensions to feet and inches: 25.19'; 68.46'; 92.10'; 145.60'; 236.21'; 33.95'.

3-20. **Conversion Table.** When distances are required to the nearest $\frac{1}{100}$ of a foot, or to the nearest $\frac{1}{8}''$, the method given above will give accurate results. If greater precision is desired, Table 4 may be used. This is a table in which inches and frac-

TABLE 4. INCHES EXPRESSED IN DECIMALS OF A FOOT

In.	0	1	2	3	4	5	6	7	8	9	10	11
0	Foot	.0833	.1667	.2500	.3333	.4167	.5000	.5833	.6667	.7500	.8333	.9167
1–32	.0026	.0859	.1693	.2526	.3359	.4193	.5026	.5859	.6693	.7526	.8359	.9193
1–16	.0052	.0885	.1719	.2552	.3385	.4219	.5052	.5885	.6719	.7552	.8385	.9219
3–32	.0078	.0911	.1745	.2578	.3411	.4245	.5078	.5911	.6745	.7578	.8411	.9245
1–8	.0104	.0938	.1771	.2604	.3438	.4271	.5104	.5938	.6771	.7604	.8438	.9271
5–32	.0130	.0964	.1797	.2630	.3464	.4297	.5130	.5964	.6797	.7630	.8464	.9297
3–16	.0156	.0990	.1823	.2656	.3490	.4323	.5156	.5990	.6823	.7656	.8490	.9323
7–32	.0182	.1016	.1849	.2682	.3516	.4349	.5182	.6016	.6849	.7682	.8516	.9349
1–4	.0208	.1042	.1875	.2708	.3542	.4375	.5208	.6042	.6875	.7708	.8542	.9375
9–32	.0234	.1068	.1901	.2734	.3568	.4401	.5234	.6068	.6901	.7734	.8568	.9401
5–16	.0260	.1094	.1927	.2760	.3594	.4427	.5260	.6094	.6927	.7760	.8594	.9427
11–32	.0286	.1120	.1953	.2786	.3620	.4453	.5286	.6120	.6953	.7786	.8620	.9453
3–8	.0313	.1146	.1979	.2813	.3646	.4479	.5313	.6146	.6979	.7813	.8646	.9479
13–32	.0339	.1172	.2005	.2839	.3672	.4505	.5339	.6172	.7005	.7839	.8672	.9505
7–16	.0365	.1198	.2031	.2865	.3698	.4531	.5365	.6198	.7031	.7865	.8698	.9531
15–32	.0391	.1224	.2057	.2891	.3724	.4557	.5391	.6224	.7057	.7891	.8724	.9557
1–2	.0417	.1250	.2083	.2917	.3750	.4583	.5417	.6250	.7083	.7917	.8750	.9583
17–32	.0443	.1276	.2109	.2943	.3776	.4609	.5443	.6276	.7109	.7943	.8776	.9609
9–16	.0469	.1302	.2135	.2969	.3802	.4635	.5469	.6302	.7135	.7969	.8802	.9635
19–32	.0495	.1328	.2161	.2995	.3828	.4661	.5495	.6328	.7161	.7995	.8828	.9661
5–8	.0521	.1354	.2188	.3021	.3854	.4688	.5521	.6354	.7188	.8021	.8854	.9688
21–32	.0547	.1380	.2214	.3047	.3880	.4714	.5547	.6380	.7214	.8047	.8880	.9714
11–16	.0573	.1406	.2240	.3073	.3906	.4740	.5573	.6406	.7240	.8073	.8906	.9740
23–32	.0599	.1432	.2266	.3099	.3932	.4766	.5599	.6432	.7266	.8099	.8932	.9766
3–4	.0625	.1458	.2292	.3125	.3958	.4792	.5625	.6458	.7292	.8125	.8958	.9792
25–32	.0651	.1484	.2318	.3151	.3984	.4818	.5651	.6484	.7318	.8151	.8984	.9818
13–16	.0677	.1510	.2344	.3177	.4010	.4844	.5677	.6510	.7344	.8177	.9010	.9844
27–32	.0703	.1536	.2370	.3203	.4036	.4870	.5703	.6536	.7370	.8203	.9036	.9870
7–8	.0729	.1563	.2396	.3229	.4063	.4896	.5729	.6563	.7396	.8229	.9063	.9896
29–32	.0755	.1589	.2422	.3255	.4089	.4922	.5755	.6589	.7422	.8255	.9089	.9922
15–16	.0781	.1615	.2448	.3281	.4115	.4948	.5781	.6615	.7448	.8281	.9115	.9948
31–32	.0807	.1641	.2474	.3307	.4141	.4974	.5807	.6641	.7474	.8307	.9141	.9974
	0	1	2	3	4	5	6	7	8	9	10	11

tions of an inch are expressed in decimals of a foot, the equivalents being carried to the fourth decimal place. Note that the fractions advance by thirty-seconds of an inch. By referring to the table, we read directly that $4\frac{17}{32}'' = 0.3776'$, $8\frac{3}{16}'' = 0.6823'$, $0.6563' = 7\frac{7}{8}''$, $0.7552' = 9\frac{1}{16}''$, etc.

CHAPTER 4

MEASURING ANGLES

4-1. **Levels and Transits.** The builders' level and transit level are simplified versions of the more accurate instruments used by engineers or professional surveyors. Fundamentally, they are used for taking elevations and angles and the basic functions of the builders' instruments are similar to those of the engineers' level and transit.

4-2. **The Builders' Level.** The essential parts of a *builders' level* are a *telescope* about 12″ in length, a graduated *horizontal circle*

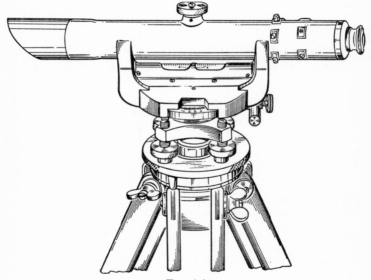

FIG. 4-1.

divided into 1-degree spaces, an attached *vernier* to permit the reading of angles to 5 minutes, a *level vial*, and four *levelling screws* by means of which the instrument is brought to a level position.

These basic parts are assembled into a single unit which is attached to a *tripod* with adjustable legs. Such an instrument is shown in Fig. 4-1. A builder's level may be revolved in only a *horizontal plane*. It is the instrument most commonly used by architects and builders.

4-3. **The Builders' Transit.** The builders' transit, Fig. 4-2, contains all the features found in the builders' level but, in addition,

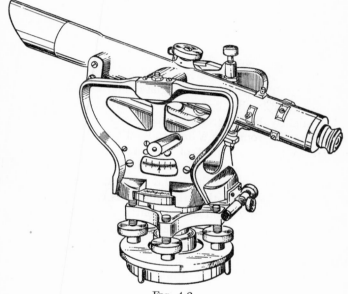

FIG. 4-2.

the telescope is capable of being revolved in a vertical as well as in a horizontal plane. To measure the vertical inclination a *vertical arc* is provided. This arc is generally divided into 1-degree spaces, and an accompanying vernier permits readings to 5 minutes. The transit permits the reading of angles of both elevation and depression to 45 degrees. In use on a sloping terrain, the transit has an obvious advantage over the level. Some builders' transits are equipped with a compass. The surveying instruments used by engineers and professional surveyors permit greater accuracy and are more versatile in their operation than the builders' levels and transits.

4-4. **The Telescope.** The telescope on levels and transits consists of metal tubes in which are found the various lenses common to the ordinary telescope. The lens at the front end is the *objective*, and the rear lens is contained in the *eye piece*. In the telescope is found a ring on which spider-web *cross hairs* are stretched at right angles to each other. Focusing on the cross hairs is accomplished by a spiral mechanism in the eye piece. Focusing on the object to be sighted is accomplished by a screw, on top of the instrument, that operates a lens located between the objective and the cross hairs; this is known as internal focusing.

4-5. **The Spirit Level.** Directly below the telescope, and parallel to it, is the *bubble tube* in which is contained a *spirit level*. See Fig. 4-1. This consists of a glass tube sealed at both ends and almost completely filled with a non-freezing fluid. The tube is either slightly bent or is a straight tube in which the upper inside surface is ground to a longitudinal circular curve. A scale is etched on the upper part of the glass tube, reading in both directions from the center. As the ends of the tube are raised or lowered the air bubble in the liquid takes various positions and, when the scale indicates that the bubble is exactly in the center of the tube, the tube, and consequently the telescope, are in a level position. For accuracy, the line of sight in the telescope and the spirit level must be parallel.

4-6. **The Horizontal Circle.** The horizontal circle on a builders' level or transit is generally graduated in intervals of 1° with each unit of 10° numbered continuously around the circle. Certain instruments have the numbering running both clockwise and counterclockwise. A thumb screw serves to clamp the telescope support to the graduated circle. By loosening the screw the telescope is permitted to revolve so that the object may be sighted. This screw is then tightened, and another thumb screw, called the *tangent screw*, is turned back and forth to permit the object to be sighted accurately.

Assume that the instrument is set up over a point, that the line of sight is on an object, and that the horizontal scale is set at 0°. By revolving the telescope and sighting on a second object the angle between the two objects may be found by reading the number of degrees on the graduated scale. Since the horizontal circle is graduated in degrees, the reading is only an approximation and a more accurate reading is obtained by the use of the vernier at the side of the horizontal circle.

4-7. The Vernier. A *vernier* is a short scale that is adjacent to the divisions of a graduated scale; its purpose is to determine the fractional part of the smallest units of the graduated scale. In Fig. 4-3, 4-4, and 4-5 a vernier is shown adjacent to a graduated

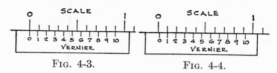

FIG. 4-3.   FIG. 4-4.

scale. On the vernier is a zero point called the *index;* the vernier is an aid in reading the position of the index on the graduated scale.

Figure 4-3 shows a portion of a $1''$ scale with subdivisions of $\frac{1}{10}''$. Below this scale is a vernier; it is $\frac{9}{10}''$ in length, and it also is divided into tenths. Consequently, each division of the vernier is $\frac{9}{10}$ of a division on the scale. In Fig. 4-3 the index on the vernier coincides with the 0 mark on the scale and hence the 1 mark on the vernier must be at a point $\frac{1}{10}$ of $\frac{1}{10}''$ (or $\frac{1}{100}''$) to the left of the

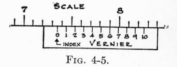

FIG. 4-5.

$\frac{1}{10}''$ mark on the scale. The 2 mark on the vernier lies at $\frac{2}{100}''$ to the left of the $\frac{2}{10}''$ mark on the scale, and so on.

Now, let us move the vernier so that the 1 mark on the vernier coincides with the $\frac{1}{10}''$ mark on the scale. This is shown in Fig. 4-4. Actually, we have moved the vernier $\frac{1}{100}''$ to the right. If we had moved the vernier $\frac{2}{100}''$ to the right, the 2 mark on the vernier would have coincided with the $\frac{2}{10}''$ mark on the scale. The vernier permits us to divide the $\frac{1}{10}''$ spaces on the scale into hundredths.

Suppose the vernier is moved to the position shown in Fig. 4-5, and we are asked to determine the exact position of the index on the vernier in relation to the adjacent $1''$ scale. We can see that it lies at some point between $7.3''$ and $7.4''$. Now, since the 6 mark on the vernier coincides with one of the divisions on the scale, the index lies $\frac{6}{100}''$ to the right of the 7.3 mark and the reading, therefore, is 7.36″.

The rule for reading a vernier of this type is: *Record the nearest scale reading adjacent to the index on the vernier. Next, observe the number of the line on the vernier that coincides with one of the division marks on the scale, and add this number to the number previously recorded.*

4-8. **Verniers Used in Measuring Angles.** The principle of the vernier, explained in Art. 4-7, is applied to reading the divisions

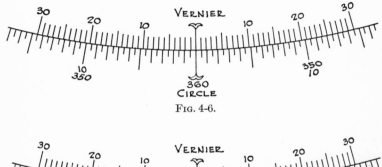

Fig. 4-6.

Fig. 4-7.

of a circular arc. The graduations of the horizontal circle and their accompanying verniers on surveying instruments are varied. Most builders' levels permit a reading to 5′, whereas, with the more accurate surveyors' instruments, the readings may be made to 1′ or even to 20″. *The smallest angle that can be read with a vernier is called the least count;* it is equal to the smallest graduation on the circle divided by the number of divisions on the vernier.

Figure 4-6 shows a vernier and a portion of a horizontal circle that are found on some transits. The index on the vernier coincides with the 0 mark on the graduated circle. Note that 30 spaces on the vernier correspond to 29 spaces on the graduated circle. The smallest division on the circle is $\frac{1}{2}$ of 1° (30′), and the number of divisions on the vernier is 30; hence, 30′ × $\frac{1}{30}$ = 1′, the least count. The vernier shown in this figure extends both to the right and to the left of the index. Such a vernier is known as a *double direct vernier;* it permits the reading of angles when the telescope is revolved in either direction. The graduations on the circle are numbered both clockwise and counterclockwise. In revolving the telescope, the outer circle remains stationary while the inner circle (the vernier) revolves with the telescope.

Figure 4-7 shows the same circle and vernier when the telescope has been revolved to another position. It is important to remember that *the arrow, or index, on the vernier points to the angle that is to be read.* Since there are two verniers, one on each side of the index, the vernier to use in reading an angle is the one that extends beyond the index in *the same direction* in which the *increasing number* of degrees is read on the circle. Since two different angles may be read, the angle desired depends upon the direction in which the telescope has been revolved. Suppose, for instance, from the position shown in Fig. 4-6, the telescope had been revolved clockwise. Then the reading on the circle would be 46 degrees plus a certain number of minutes. Then, reading the vernier in the same direction (to the left of the index), we find that the line on the vernier that coincides with a line on the graduated circle is line 21. Therefore, the reading is 46 degrees plus 21 minutes, 46° 21′.

Now suppose that the telescope had been revolved counterclockwise. The index points to 313 degrees 30 minutes plus some more minutes. Reading the vernier to the right of the index we see that the 9 mark on the vernier coincides with a graduation on the circle, and therefore the reading is 313° 30′ plus 9″, or 313° 39′. The sum of the two readings should equal 360°. Thus 46° 21′ + 313° 39′ = 360°.

Figure 4-8 shows a circle and double vernier often found on a builders' transit. Note that the circle is divided into degrees (units of 60′) and that 12 spaces on the vernier equal 11 spaces on the graduated circle. This indicates that the least count is 60′ × $\frac{1}{12}$, or 5′. Therefore, the smallest angle that can be read is to 5′.

Figure 4-9 shows the same circle and vernier when the telescope has been revolved to a new position. If the telescope has been turned clockwise the index points to a reading of 66 degrees plus some minutes. Reading in the same direction (to the left of the index), we see that the fourth space on the vernier coincides

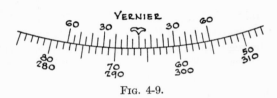

FIG. 4-8.

FIG. 4-9.

with a space on the graduated circle. Therefore, since the least count is 5′, 4 × 5 = 20′; hence the reading is 66° 20′. In a similar manner, if the telescope had been turned counterclockwise, the angle is read 293° 40′.

4-9. **The Compass.** A part of an engineers' transit is the compass; it is also found on some builders' transits. The magnetic needle of the compass automatically points to the *north magnetic pole*. This is not the *true geographic pole*, and the angle between them is called the *magnetic declination*. The declination for a particular location may be found on government charts as, for example, 10° W, by which we mean that the compass points 10° west of the true or geographic north. All bearings given on surveys or site plans must be related to *the true north;* hence, whenever possible, the bearings on surveys should be related to some established line the true bearing of which is known.

The pivot of the compass is at the center of a horizontal circle that is divided into quadrants, the quadrants being graduated into 90° each. To give a *bearing* of a certain point or line, it is necessary to give both the quadrant and the number of degrees. For example, a bearing of N 35° W indicates that the direction of an object sighted is on a line that is 35° west of north. See Art. 5-3.

4-10. Setting Up the Instrument. When the level or transit is to be set up over a given point, extreme care must be taken. Having attached the instrument to the tripod, the tripod is forced firmly into the ground so that the instrument is approximately over the point in the stake. The four levelling screws are operated so that the spirit level shows the instrument to be level. The plumb bob suspended from the instrument shows the relation of the instrument to the point in the stake, and a slight loosening of the levelling screws permits the head to be moved so that the plumb bob is brought directly over the point. When this has been accomplished the levelling screws are again operated until the spirit level indicates the instrument to be level in all directions.

4-11. Measuring Horizontal Angles. Suppose that the instrument has been set up over a given point and that we wish to measure the angle between two distant points, the point of the set-up being the vertex of the angle. The rodman goes to one of the points and holds a ranging pole in a vertical position directly over the point. The clamp screw on the instrument is loosened and the telescope is revolved so that the pole and cross hairs are in approximate alignment. The clamp screw is now tightened, and the tangent screw is operated to bring the pole and cross hairs into an exact line. The angle on the horizontal circle is now read and recorded.

The rodman now moves to the other point, the instrument is sighted on this point, and the angle is again read and recorded. The difference between the two readings is the angle between the two points. When a transit is used instead of a level, the instrument may be sighted directly on the points, eliminating the necessity of the ranging pole and insuring greater accuracy. Many instruments permit the setting of 0 on the first point, thus obtaining the angle directly.

When the construction of the surveying instrument permits, angles may be measured with greater accuracy by using the *method of repetition.* Referring to Fig. 4-10, suppose we are required to measure the angle *BAC*. The procedure is as follows:

STEP 1. Set up the instrument over point *A* with both clamps loose, and set 0 of the graduated circle and the index of the vernier approximately together. Tighten the *upper clamp*, and

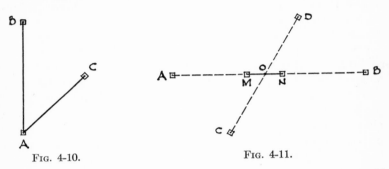

FIG. 4-10.   FIG. 4-11.

with the upper tangent screw bring the index and the 0 of the graduated circle together.

STEP 2. With the upper clamp tight and the lower clamp loose, turn the instrument to sight approximately on point *B*. Tighten the *lower clamp*, and, with the lower tangent screw, set the vertical cross hair exactly on point *B*. The 0 of the instrument is now set on *B*.

STEP 3. Loosen the *upper clamp*, and turn the telescope to sight on point *C*. Tighten the upper clamp, and, with the upper tangent screw, bring the vertical cross hair to coincide with point *C*. Record this reading.

STEP 4. Now repeat Step 2. The angular value of *BAC* is now sighted on point *B*.

STEP 5. Repeat Step 3. The reading now taken is twice the actual angle *BAC*. By dividing this reading by 2 we obtain the *average* of the two readings which, of course, is more accurate than the first reading taken in Step 3.

4-12. **Setting Points on Line.** When two points have been established, other points on the line between them may be readily determined. The instrument is set up and levelled over one of the given points and sighted on the other point. The line of sight of the telescope is now in a vertical plane that passes through the two established points. To locate any other point on this line the rodman goes to the approximate position and moves the ranging pole to the right or left (as directed by the instrument man) until the image of the pole and the cross hairs are in alignment. The rodman drives a stake in the ground at this point. For accuracy, the operation is repeated and the rodman marks with a tack the point on the stake that is in exact alignment with the two given points.

4-13. **Laying Off an Angle.** Suppose that a line has been established and that it is required to lay off an angle of a given magnitude from some point on the line. First, the instrument is set up and levelled over the given point and the instrument is sighted on the established line. The angular reading on the horizontal circle is now recorded. To this angle we add (or subtract, as may be required) the given angle. The clamp screw on the instrument is loosened, and the telescope is turned until the index of the vernier points (approximately) to the sum (or difference) of the angles. Now the clamp screw is tightened, and the tangent screw is operated so that the proper line on the vernier coincides with a line on the graduated circle. The telescope has now been revolved the required angle, and a stake and tack are located on this line, thus establishing the required angle.

The method of laying off angles by repetition is explained in Art. 16-1.

4-14. **Intersection of Lines.** A problem that frequently occurs in staking out buildings is to find the point of intersection of two previously established lines. Referring to Fig. 4-11, assume that points A, B, C, and D have been established and that we wish to find the point of intersection of lines AB and CD. First, set up the instrument over point A and sight on point B. The line of sight in the telescope now lies in a vertical plane that includes line AB. Stakes are now driven on line AB at points M and N. Points M and N are several feet apart and are located on the left and right sides of line CD, respectively. By means of the plumb bob, points M and N are accurately established in the stakes by tacks and a string is stretched between them. Now move to position C, set up the instrument, and sight on point D. This line of sight intersects the string at point O, at which point a stake is driven and, with the aid of a plumb bob, a tack is driven to indicate the intersection of lines AB and CD.

4-15. **Suggestions.** In a book of this scope detailed instructions for setting up and operating a level or transit are impracticable. The best method of acquiring facility in its use is to obtain an instrument, follow closely the accompanying instructions, and practice the various procedures. The following suggestions may prove to be helpful.

In setting up the instrument be certain that the legs of the tripod are firmly forced into the ground so that the instrument is not easily disturbed. Make certain that the cross hairs in the telescope are in sharp focus. Be sure that the plumb bob is exactly centered over the tack in the stake. While using the instrument make frequent checks on the spirit level to see that the instrument has not been disturbed. When using a double vernier be sure that the reading is made on the proper side of the index. If the horizontal circle has two rows of numbers in opposite directions be careful to see that the reading is taken in the proper direction. In taking an angular reading of a circle, as illustrated in Fig. 4-7, do not forget to add the 30'; the reading is $313° + 30'$ $+ 9'$, $313° 39'$, not $313° 9'$. Follow carefully the manufacturer's instructions relating to the care of the instrument.

CHAPTER 5

LAND SURVEYS

5-1. Surveys. Because of legal aspects, the original survey used for writing deed descriptions, or for staking out the boundary lines of a property, should be entrusted only to a registered surveyor. The architect or builder, before proceeding with any design or construction work, should require that the owner furnish this certified survey. Additional lines and grades may need to be determined within the boundary lines of the plot in order to locate buildings, roads, paths, etc. This work may be performed by the architect or builder if he is qualified. The architect does not make the original survey. However, in an illustrative example, consideration of which runs intermittently through this and the succeeding chapters, beginning in Art. 5-5, the complete computations are given to explain the procedure and the various problems that are generally encountered. After the data concerning angles, length of lines, etc., have been obtained in the field, the necessary computations are made. Many methods may be employed in obtaining data, but the computations given in the illustrative example apply to all survey work.

Plane surveying treats the surface of the earth as a plane surface. Although this is not theoretically exact, the assumption is sufficiently accurate when surveys of comparatively small areas are involved.

5-2. Traverses. *A traverse* is a line or a series of connected lines surveyed across the earth's surface. *An open traverse* begins at a given point and ends at some distant point, as, for example, a survey of a highway or railroad. *A closed traverse* begins at a given point and returns to the same point, thus forming a closed circuit. This closed type of traverse is applicable to surveys of parcels of land, the boundaries forming a polygon.

5-3. Bearings of Lines. The bearing of a line is the horizontal angle between the direction of the line and a line pointing to the

true north. For example, we say the bearing of a certain line is
N 36° E. This indicates that the line is measured from the north
in an easterly direction at an angle of 36°, as shown in Fig. 5-1.
This bearing might also have been determined by measuring from
the south, in which case the bearing would be recorded S 36° W.
Bearings are always measured from either the north or south and

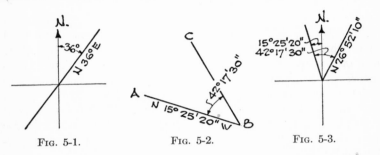

<table>
<tr><td>Fig. 5-1.</td><td>Fig. 5-2.</td><td>Fig. 5-3.</td></tr>
</table>

from no other cardinal point, the angular value never being
greater than 90°.

A practical method of relating the lines on a survey to the true
north is to relate one of the lines to some line whose bearing has
already been established. Such lines may be obtained from highway and city plans. If such a line forms a portion of the boundary
of the plot, the bearings of the remaining lines are readily determined.

EXAMPLE

EXAMPLE. In Fig. 5-2 line AB has a bearing of N 15° 25′ 20″ W.
Line BC intersects AB at an angle of 42° 17′ 30″, as shown. Determine the bearing of line BC.

SOLUTION. For problems of this kind *always make a sketch* in
which both lines are related to north, as shown in Fig. 5-3. By
examining the sketch it is apparent that BC lies east of north and
its bearing is the difference between 42° 17′ 30″ and 15° 25′ 20″,
or N 26° 52′ 10″ E. The bearing of this line might also be given
as S 26° 52′ 10″ W. The bearing is not always the difference between the two angles. It is necessary that the sketch be made,
for the sketch indicates the proper procedure.

PROBLEMS

5-3-A-B-C-D-E-F. In each of the diagrams shown in Fig. 5-4 two lines and the angles between them are shown, the bearing of one line being given. Compute the bearings of the remaining lines.

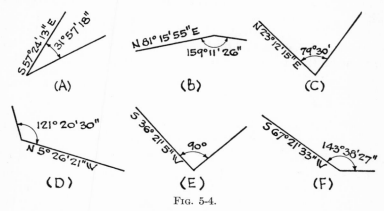

FIG. 5-4.

5-4. Intersecting Lines. Two non-parallel lines that intersect form four angles, as shown in Fig. 5-5. The opposite angles are equal; angle 1 = angle 3 and angle 2 = angle 4. Any two of the

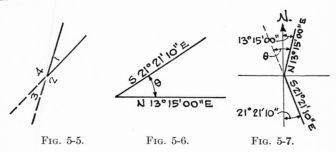

FIG. 5-5.   FIG. 5-6.   FIG. 5-7.

adjacent angles are *supplementary*, that is, their sum is equal to 180°.

We frequently have a problem in which the bearings of two lines are given, and we are required to find the angle between them. *Always* make a sketch, and note which angle is required. Do not confuse the required angle with its supplement. Is the required angle acute or obtuse, less or greater than 90°?

EXAMPLES

EXAMPLE. Two lines whose bearings are given are shown in Fig. 5-6. Determine θ, the angle of intersection.

SOLUTION. A sketch is made, showing the bearings of the lines with relation to the points of the compass, Fig. 5-7. From Fig. 5-6 we see that θ, the required angle, is the acute angle. Therefore, in Fig. 5-7, we extend the line whose bearing is S 21° 21′ 10″ E and we see that to obtain the acute angle we must add the given angles. Hence

$$21° \ 21′ \ 10″$$

$$\text{plus } 13° \ 15′ \ 00″$$

$$\overline{\rule{4cm}{0pt}}$$

$$34° \ 36′ \ 10″ = \theta \quad \text{the required angle}$$

EXAMPLE. Two intersecting lines whose bearings are given are shown in Fig. 5-8. Determine angle θ.

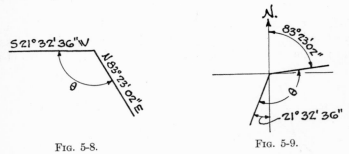

FIG. 5-8.   FIG. 5-9.

SOLUTION. Figure 5-9 is the sketch showing the positions of the lines and their angles in relation to the points of the compass. From Fig. 5-8 note that θ, the required angle, is the obtuse angle. Then

$$180° \ 00′ \ 00″$$

$$\text{plus} \quad 21° \ 32′ \ 36″$$

$$\overline{\rule{4cm}{0pt}}$$

$$201° \ 32′ \ 36″$$

$$\text{minus} \quad 83° \ 23′ \ 02″$$

$$\overline{\rule{4cm}{0pt}}$$

$$118° \ 09′ \ 34″ = \theta \quad \text{the required angle}$$

PROBLEMS

5-4-A-B-C-D-E-F. In each of the diagrams shown in Fig. 5-10 two intersecting lines and their bearings are given. For each pair of lines, compute
angle θ.

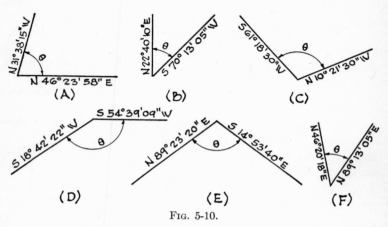

FIG. 5-10.

5-5. Making a Survey. To make a survey of an existing plot of
ground the corners must first be definitely located and marked.
If possible, the instrument should be set up directly over these
points. In the following illustrative example it is assumed that

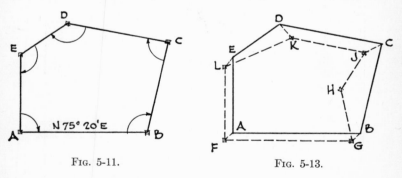

FIG. 5-11.   FIG. 5-13.

this can be done. The method of making the survey described
here is but one of several; it is particularly applicable when the
builders' level or transit is used. Let us assume that the plot to
be surveyed has the corners located as shown in Fig. 5-11. In

moving the instrument from point to point we will take the points in this sequence: *A, B, C, D, E*. The angles measured will be the interior angles, and the instrument will always be turned to the right (clockwise direction) in reading the angles. The bearing of line *AB*, N 75° 20′ E, has been obtained from the township surveyor. To make the survey, the following steps are taken:

STEP 1. Set up the instrument over any point, say point *A*, and sight on point *E*.

STEP 2. Record the angular reading on the horizontal graduated circle.

STEP 3. Revolve the instrument (clockwise) and sight on point *B*.

STEP 4. Take the angular reading. The difference between the two recorded angles will be the internal angle at *A*.

STEP 5. Measure distance *AE*. For greater accuracy the distance should be measured twice, first from *A* to *E* and then from *E* to *A*. If there is only a slight difference in the measurements, their average may be taken as the length of the line.

SURVEY OF JOHN E. DOE PROPERTY
NEWVILLE, BUCKS CO., PA.

SURVEY PARTY
ABLE
BAKER
COOPER

DATE — AUG. 21, 1953

ON STATION	SIGHT AT	ANGULAR READING	INTERNAL ANGLE	DISTANCE	AVERAGE DISTANCE
A	E	177° 15′	90° 00′	62.58	62.60
	B	267° 15′	30′	101.70	
B	A	126° 20′	102° 25′	102.10	101.90
	C	228° 45′		75.03	
C	B	321° 15′	86° 40	74.91	74.97
	D	47° 55′	20′	84.41	
D	C	72° 20′	136° 15′	84.55	84.48
	E	208° 35′		42.70	
E	D	102° 00′	124° 30′	42.68	42.69
	A	226° 30′		62.62	

$\Sigma \angle s = 539° 50′$

$180 (n-2) = 180 \times 3 = 540°$

FIG. 5-12.

STEP 6. The instrument is now moved to point *B*, sights are taken on points *A* and *C*, and the above procedure is repeated.

This process is repeated at each successive point so that all the interior angles and the lengths of lines have been measured. For greater accuracy the entire procedure may be repeated and the measurements of the angles averaged.

Figure 5-12 shows a suggested form for recording the notes of the survey. Field books with ruled pages may be obtained for this purpose. In some books the left-hand page is used for the data shown in Fig. 5-12 and the right-hand page is used for sketches and notes.

The survey shown in Fig. 5-12 was made with an instrument having the horizontal circle graduated and marked from 0° to 360°. Some instruments are graduated in 90° quadrants and are so constructed that the horizontal circle may be set at 0, thereby eliminating the first reading and the subsequent subtraction necessary to obtain the angle.

5-6. **Checking Interior Angles.** Before leaving the site, it is well to make an additional check on the accuracy of the angles. From plane geometry we know that in any closed polygon the sum of the interior angles equals $180° \times (n - 2)$, where n is the number of sides of the polygon. In this example the polygon has five sides and therefore the sum of the interior angles should equal $180 \times (5 - 2) = 180 \times 3$, or 540°.

Note, in Fig. 5-12, that the sum of the angles does not equal 540°; it is 539° 50′, a difference of 10′. This is usually the case. When using the builders' level or transit, which has an accuracy of only 05′ in reading angles, the discrepancy should not exceed $05'\sqrt{n}$. In this instance $n = 5$; hence $05' \times \sqrt{5} = 11.18'$. Since the difference of 10′ falls within the allowable limit, this indicates acceptable work considering the accuracy of the instrument.

5-7. **Balancing Interior Angles.** The discrepancy of 10′, explained in Art. 5-6, must be distributed among the five angles. It might appear logical to distribute the 10′ equally among each of the angles, adding 02′ to each. This would result in angles such as 86° 42′, implying that the survey had been made with an instrument measuring to 01′. This, however, is known to be untrue; hence we will make the corrections by adding 05′ to each of the angles at B and D, as shown in Fig. 5-12. These two angles have been chosen arbitrarily. An assumption often made is that the angles with the shortest sides will have the greatest errors. If conditions in the field indicate that an error may have been made at some other angle or angles, the correction should be made at these points.

5-8. Alternative Positions of the Instrument. When the condition arises where obstructions prevent placing the instrument directly over the corners of the plot, stakes are driven at adjacent points, F, G, H, J, K, and L, as indicated in Fig. 5-13. These points are selected to permit an unobstructed view of each other as well as of the corners of the plot. A closed traverse is made of these points, and lines FA, GB, JC, etc., are run to the corner points. These lines and the angles they make with the traverse are measured. With this information the coordinates of points A, B, C, D, and E may be determined (see Art. 6-9) and the boundaries of the plot thus established.

CHAPTER 6

SURVEY COMPUTATIONS

6-1. **Survey Computations.** Chapter 5 explained the procedure of determining the angles and the lengths of lines in surveying a plot of ground. These data are obtained at the site. The next step is to perform certain computations that enable one to plot the survey and to determine its area. The computations are generally made in the office.

6-2. **Plotting the Survey.** From the data found in the field, shown in Fig. 5-12, the plot is drawn to scale, Fig. 6-1. This

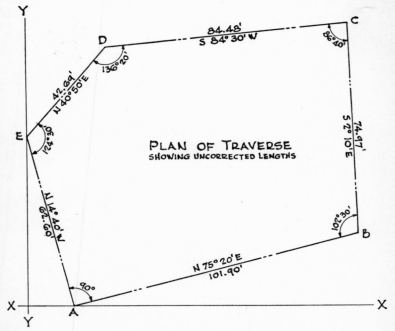

PLAN OF TRAVERSE
SHOWING UNCORRECTED LENGTHS

Fig. 6-1.
71

diagram shows the internal angles, and the lengths and the bearings of all lines. The bearings are computed as explained in Art. 5-3. From the township surveyor we found that the bearing of *AB* is N 75° 20′ E. The angle between *AB* and *BC* is 102° 30′; hence the bearing of *BC* is S 2° 10′ E. The bearings of the other sides are likewise determined. A protractor may be used in laying off the angles, but more accurate methods of plotting are explained in Art. 6-10. Note, in Fig. 6-1, that the lines have been drawn in relation to true north as a vertical axis. This should always be done since it simplifies computations.

6-3. **Latitudes and Departures.** The *latitude* of a line is its projection on the north and south line. The *departure* of a line is its projection on the east and west line. They are the vertical and horizontal coordinates, as shown in Fig. 6-2. In this figure *AB is the given line and its bearing is the angle θ*. Note that the angle at *B* is also angle *θ*. The length of line *AB*

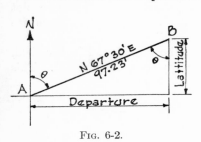

Fig. 6-2.

is 97.23′. In the right triangle observe the following relationships:

$$\text{departure} = \text{length} \times \sin \text{bearing}$$

$$\text{latitude} = \text{length} \times \cos \text{bearing}$$

By making sketches, it is seen that these relationships are valid regardless of the quadrant in which the line is drawn.

For line *AB* shown in Fig. 6-1,

$$\text{departure} = 97.23 \times \sin 67° 30′ = 89.83′$$

$$\text{latitude} = 97.23 \times \cos 67° 30′ = 37.21′$$

log 97.23	=	1.98780	log 97.23	=	1.98780
log sin 67° 30′	=	9.96562 −10	log cos 67° 30′ =		9.58284 −10
		1.95342			1.57064
		= 89.83′			= 37.21′

The latitude and departure of all the sides of the plot shown in Fig. 6-1 are computed as shown in Fig. 6-3. After they are computed, the latitudes and departures can be checked approximately by scaling the corresponding distances in Fig. 6-1. This check will disclose any large errors in computations.

Departures   Latitudes

LINE AB

log 101.90   = 2.00817   log 101.90   = 2.00817
log sin 75°20' = 9.98561   log cos 75°20' = 9.40346
              1.99378   1.41163
            = 98.58   = 25.80

LINE BC

log 74.97   = 1.87489   log 74.97   = 1.87489
log sin 2°10' = 8.57757   log cos 2°10' = 9.99969
             0.45246   1.87458
           = 2.834   = 74.92

LINE CD

log 84.48   = 1.92675   log 84.48   = 1.92675
log sin 84°30' = 9.99800   log cos 84°30' = 8.98157
              1.92475   0.90832
            = 84.09   = 8.097

LINE DE

log 42.69   = 1.63033   log 42.69   = 1.63033
log sin 40°50' = 9.81549   log cos 40°50' = 9.87887
              1.44582   1.50920
            = 27.91   = 32.30

LINE EA

log 62.60   = 1.79657   log 62.60   = 1.79657
log sin 14°40' = 9.40346   log cos 14°40' = 9.98561
              1.20003   1.78218
            = 15.85   = 60.56

FIG. 6-3.

In traversing a route on a plane surface, and returning to the starting point, a closed traverse, we must travel the same distance north as we do south and just as far east as we do west. Let us consider the distances travelled north and south as plus latitudes and minus latitudes, respectively, and distances travelled east and west as plus departures and minus departures. Then, *in any closed polygon or traverse, the plus latitudes must equal the minus latitudes and the plus departures must equal the minus departures.*

The latitudes and departures computed in Fig. 6-3 are tabulated in Fig. 6-4 with respect to their plus and minus signs. Note, in Fig. 6-1, that in going from A to B we travel northward and eastward; therefore both latitude and departure are placed in the plus columns. From B to C we go northward and westward, a plus latitude and minus departure. The other values are plotted

LINE	LATITUDES		DEPARTURES		BALANCED LATITUDES	BALANCED DEPARTURES
	+	−	+	−		
AB	25.80		98.58		25.83	98.75
BC	74.92			2.83	75.01	2.83
CD		8.10		84.09	8.09	83.94
DE		32.30		27.91	32.26	27.86
EA		60.56	15.85		60.49	15.88

$$
\begin{array}{cccc}
100.72 & 100.96 & 114.43 & 114.83 \\
 & 100.72 & & 114.43 \\
 & .24 & & .40
\end{array}
$$

$$\text{Error of Closure} = \sqrt{(.24)^2 + (.40)^2} = .47$$

$$\text{Precision} = \frac{.47}{366.64} = \frac{1}{780}$$

CORRECTIONS

	LAT.	DEP.
AB	.03	.17
BC	.09	.00
CD	.01	.15
DE	.04	.05
EA	.07	.03
	.24	.40

FIG. 6-4.

similarly. We find, in Fig. 6-4, that the sum of the plus latitudes does not equal the sum of the minus latitudes. This is true also with respect to the departures; it indicates that the polygon does not close. It is improbable that these plus and minus values will balance in any survey. However, in order to have a polygon that closes mathematically, the plus and minus columns must be balanced and the lengths of the lines correspondingly adjusted.

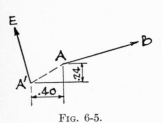

FIG. 6-5.

6-4. **Error of Closure.** If we had accurately plotted the traverse shown in Fig. 6-1 to a large scale, beginning at A and continuing in a counterclockwise direction, we would find that the end of the last line would not coincide with point A from which we started. We would have had the condition shown in Fig. 6-5. The distance between the points A' and A, the length of line by which the polygon fails to close, is called the *error of closure*. Since the differences in the plus and minus latitudes and departures are shown in Fig. 6-4 to be 0.24 and 0.40, respectively, the error of closure is the hypothenuse $(A'A)$ of the right triangle of which 0.24 and 0.40 are the other two sides, as shown in Fig. 6-5. Then

$$\text{hypothenuse} = \sqrt{0.24^2 + 0.40^2} = \sqrt{0.2176}$$
$$= 0.47' \quad \text{the error of closure}$$

6-5. Precision. Some error of closure is to be expected but, assuming there is no mistake in the computations, a large error indicates that a mistake has occurred in making the survey. The degree of error varies with the length of the lines and the accuracy of the instrument that is used.

The *precision of the survey* is equal to the error of closure divided by the sum of the lengths of the lines as expressed by a fraction with unity in the numerator. In our survey,

$$\text{precision} = \frac{0.47}{366.64} = \frac{1}{780}$$

The precision to be expected in a survey made with a builders' level or transit, measuring to the nearest 05', is about 1/500. In our survey the precision is somewhat greater, indicating that the work was performed acceptably. A precision of 1/500 is suitable for farm surveying, and it is sufficiently accurate for laying out buildings, roads, etc., within the plot. In city surveying greater precision is required and precisions of 1/10,000 and 1/20,000 are often obtained.

6-6. Corrections to Latitudes and Departures. Since the latitudes and departures must be balanced, corrections must be made. The corrections in latitude and departure are distributed proportionately among *all sides* of the survey. The correction to be applied to the latitude of each side will be

$$\frac{\text{total error in latitude} \times \text{latitude of the side}}{\text{sum of the latitudes of all sides}}$$

Therefore, the correction to be applied to the latitude of line AB will be

$$\frac{0.24 \times 25.80}{201.68} = 0.0307 \quad \text{say } 0.03'$$

The correction need only to be taken to the nearest $\frac{1}{100}$ of a foot.

The corrections to the departures are found in a similar manner. The corrections to be applied to the departure of each side will be

$$\frac{\text{total error in departure} \times \text{departure of the side}}{\text{sum of the departures of all sides}}$$

Hence, the correction to be applied to the departure of line AB will be

$$\frac{0.40 \times 98.58}{229.26} = 0.173 \quad \text{say } 0.17'$$

The corrections to latitudes and departures for all the sides of the survey are computed similarly.

6-7. **Balancing Latitudes and Departures.** The magnitudes of the corrections having been computed, they must be applied to the latitudes and departures shown in Fig. 6-4. In applying the corrections, *subtract the corrections from the values in the columns having the greater sums* and *add the corrections to the values in the columns having the smaller sums.* This procedure has been followed and the corrected (balanced) values are shown in the last two columns of Fig. 6-4. By checking, it is seen that the plus and minus latitudes and departures now balance.

6-8. **Corrected Lengths and Bearings.** The changes in the magnitudes of the latitudes and departures necessitate slight changes in the lengths of the sides of the survey and, possibly, changes in their bearings. These adjusted lengths are now computed by use of these two equations:

$$\text{length} = \sqrt{\text{latitude}^2 + \text{departure}^2}$$

$$\text{tan bearing} = \frac{\text{departure}}{\text{latitude}}$$

To verify these relationships, see Fig. 6-2.

The revised lengths are shown in the following computations, but the complete logarithmic work is shown only for line AB.

Line AB

$$\text{length} = \sqrt{\text{latitude}^2 + \text{departure}^2}$$

$$\text{length} = \sqrt{25.83^2 + 98.75^2} = 102.07'$$

$$\text{tan bearing} = \frac{\text{departure}}{\text{latitude}} = \frac{98.75}{25.83}$$

$$\text{bearing} = \text{N } 75° \ 20' \text{ E}$$

$$\log^2 25.83 = 1.41212 \times 2 = 2.82424$$
$$\log^2 98.75 = 1.99454 \times 2 = 3.98908$$

$$= \quad 667.2$$
$$= \quad 9,752$$
$$\overline{\qquad\quad 10,419}$$

$$\log 10,419 = 4.01783$$
$$\div 2$$
$$\overline{\qquad 2.00892}$$

$$.00860$$
$$\overline{\qquad 32} \qquad \tfrac{32}{43} = .7 = 102.07'$$

$$\log 98.75 = 1.99454$$
$$\text{colog } 25.83 = 8.58788 \ -10$$
$$\overline{\qquad 0.58242}$$

$$\text{bearing} = \text{N } 75° \ 20' \text{ E}$$

Line BC

$$\text{length} = \sqrt{75.01^2 + 2.83^2} = 75.07'$$

$$\text{tan bearing} = \frac{2.83}{75.01}$$

$$\text{bearing} = \text{S } 2° \ 10' \text{ E}$$

Line *CD*

$$\text{length} = \sqrt{8.09^2 + 83.94^2} = 84.33'$$

$$\tan \text{bearing} = \frac{83.94}{8.09}$$

$$\text{bearing} = \text{S } 84° 30 \text{ W}$$

Line *DE*

$$\text{length} = \sqrt{32.26^2 + 27.86^2} = 42.63'$$

$$\tan \text{bearing} = \frac{27.86}{32.26}$$

$$\text{bearing} = \text{N } 40° 50' E$$

Line *EA*

$$\text{length} = \sqrt{60.49^2 + 15.88^2} = 62.54'$$

$$\tan \text{bearing} = \frac{15.88}{60.49}$$

$$\text{bearing} = \text{N } 14° 40' \text{ W}$$

The corrected lengths of the sides are shown in Fig. 8-1. In the above computations note that the bearings did not change by an amount sufficient to affect the bearing to the nearest 05'. This will generally be the case when the error of closure is within the limit previously noted. In these computations we have gone around the traverse in a counterclockwise direction because the original survey notes were designated in this manner. A clockwise direction might have been taken; the results would have been the same.

6-9. **Coordinates of the Survey.** If any other lines or points within the plot are to be determined, or if the area of the plot is to be computed, it is necessary to know the X and Y coordinates of all the corners of the survey.

Referring to Fig. 6-6, the X coordinate of point A is its distance from the Y–Y axis; this is identified as distance X_A, and it is called the *abscissa*. Similarly, the Y coordinate of point A is Y_A, the *ordinate*.

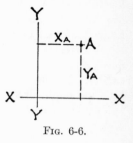

Fig. 6-6.

The X and Y coordinates of the corners of the survey are shown in Fig. 6-7. On referring to Fig. 6-1, it is seen that the X coordinate of point A is equal to the departure of line EA; the X coordinate of point B is the departure of line EA plus the departure of line AB, and so on. It should be noted that the X–X and Y–Y axes for this survey have been arbitrarily chosen to pass through the most westerly and most southerly points of the perimeter of the survey.

	X-COORDINATES	Y-COORDINATES
A =	15.88	0
	+ 98.75	25.83
B =	114.63	25.83
	− 2.83	+ 75.01
C =	111.80	100.84
	− 83.94	− 8.09
D =	27.86	92.75
	− 27.86	− 32.26
E =	0	60.49

Fig. 6-7.

6-10. **Plotting the Survey.** After the lengths of the boundaries of the plot have been determined, as well as the angles between the intersecting sides, these data must be plotted on the drawing board. To accomplish this several methods are available.

(*a*) THE PROTRACTOR. To employ this method the angles between the intersecting sides of the boundary are laid off in accordance with the divisions, indicating the degrees, marked on the protractor. The lengths of the sides are then laid off to scale on the appropriate sides. If great precision is required it cannot be obtained by this procedure. The *adjustable triangle* is sometimes used for laying off angles, but it too does not afford a great degree of accuracy. The use of a *vernier protractor* permits greater accuracy than that obtained with the ordinary protractor.

(b) PLOTTING COORDINATES. When the X and Y coordinates of the corners of the survey have been computed they may be accurately plotted to the desired scale and the lines of the survey are thus established. For this work make certain that the triangle used has a *true right angle*.

(c) PLOTTING BY TANGENTS. The method of plotting by tangents affords a great degree of accuracy and is simple in its application. It requires a scale divided into inches and tenths of an inch (an engineer's scale) and a table of *natural tangents*. Table 5 is a table giving the natural tangents of angles up to and including 45°.

EXAMPLES

EXAMPLE. Line AB in Fig. 6-8 is horizontal. Let it be required to lay off an angle of 23° 20′, the angle BAD.

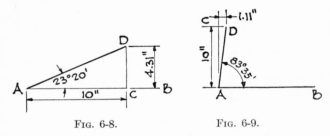

FIG. 6-8.   FIG. 6-9.

SOLUTION. On referring to Table 5 we find that the natural tangent of 23° 20′ is 0.431. Now lay off AC, a length of 10″. At point C erect a vertical line and on this line measure a distance 10×0.431 or 4.31″; call it point D. Then, since tan angle $DAC = DC/AC$, or 4.31/10, the line from A to D makes an angle of 23° 20′ with line AB. For greater accuracy, AC might have been made 20″, in which case CD would be 20×0.43136, or 8.63″.

EXAMPLE. In Fig. 6-9 the line AB is horizontal. Let it be required to lay off an angle of 83° 35′, the angle BAD.

SOLUTION. From a table of natural tangents we find the tangent of 83° 35′ to be 8.8934. By the procedure used in the previous example, the vertical line would be 10×8.8934, or 88.934″. Obviously, this length is too great to measure on the drawing board. Therefore, *when the angle to be laid off exceeds 45°, lay off the complement of the angle.* The complement of 83° 35′ = 90° 0′ −

TABLE 5. NATURAL TANGENTS

Angle	0'	10'	20'	30'	40'	50'	60'
0°	0.00000	0.00291	0.00582	0.00873	0.01164	0.01455	0.01746
1	0.01746	0.02036	0.02328	0.02619	0.02910	0.03201	0.03492
2	0.03492	0.03783	0.04075	0.04366	0.04658	0.04949	0.05241
3	0.05241	0.05533	0.05824	0.06116	0.06408	0.06700	0.06993
4	0.06993	0.07285	0.07578	0.07870	0.08163	0.08456	0.08749
5	0.08749	0.09042	0.09335	0.09629	0.09923	0.10216	0.10510
6	0.10510	0.10805	0.11099	0.11394	0.11688	0.11983	0.12278
7	0.12278	0.12574	0.12869	0.13165	0.13461	0.13758	0.14054
8	0.14054	0.14351	0.14648	0.14945	0.15243	0.15540	0.15838
9	0.15838	0.16137	0.16435	0.16734	0.17033	0.17333	0.17633
10	0.17633	0.17933	0.18233	0.18534	0.18835	0.19136	0.19438
11	0.19438	0.19740	0.20042	0.20345	0.20648	0.20952	0.21256
12	0.21256	0.21560	0.21864	0.22169	0.22475	0.22781	0.23087
13	0.23087	0.23393	0.23700	0.24008	0.24316	0.24624	0.24933
14	0.24933	0.25242	0.25552	0.25862	0.26172	0.26483	0.26795
15	0.26795	0.27107	0.27419	0.27732	0.28046	0.28360	0.28675
16	0.28675	0.28990	0.29305	0.29621	0.29938	0.30255	0.30573
17	0.30573	0.30891	0.31210	0.31530	0.31850	0.32171	0.32492
18	0.32492	0.32814	0.33136	0.33460	0.33783	0.34108	0.34433
19	0.34433	0.34758	0.35085	0.35421	0.35740	0.36068	0.36397
20	0.36397	0.36727	0.37057	0.37388	0.37720	0.38053	0.38386
21	0.38386	0.38721	0.39055	0.39391	0.39727	0.40065	0.40403
22	0.40403	0.40741	0.41081	0.41421	0.41763	0.42105	0.42447
23	0.42447	0.42791	0.43136	0.43481	0.43828	0.44175	0.44523
24	0.44523	0.44872	0.45222	0.45573	0.45924	0.46277	0.46631
25	0.46631	0.46985	0.47341	0.47698	0.48055	0.48414	0.48773
26	0.48773	0.49134	0.49495	0.49858	0.50222	0.50587	0.50953
27	0.50953	0.51320	0.51688	0.52057	0.52427	0.52798	0.53171
28	0.53171	0.53545	0.53920	0.54296	0.54673	0.55051	0.55431
29	0.55431	0.55812	0.56194	0.56577	0.56962	0.57348	0.57735
30	0.57735	0.58124	0.58513	0.58905	0.59297	0.59691	0.60086
31	0.60086	0.60483	0.60881	0.61280	0.61681	0.62083	0.62487
32	0.62487	0.62892	0.63299	0.63707	0.64117	0.64528	0.64941
33	0.64941	0.65355	0.65771	0.66189	0.66608	0.67028	0.67451
34	0.67451	0.67875	0.68301	0.68728	0.69157	0.69588	0.70021
35	0.70021	0.70455	0.70891	0.71329	0.71769	0.72211	0.72654
36	0.72654	0.73100	0.73547	0.73996	0.74447	0.74900	0.75355
37	0.75355	0.75812	0.76272	0.76733	0.77196	0.77661	0.78129
38	0.78129	0.78598	0.79070	0.79544	0.80020	0.80498	0.80978
39	0.80978	0.81461	0.81946	0.82434	0.82923	0.83415	0.83910
40	0.83910	0.84407	0.84906	0.85408	0.85912	0.86419	0.86929
41	0.86929	0.87441	0.87955	0.88473	0.88992	0.89515	0.90040
42	0.90040	0.90569	0.91099	0.91633	0.92170	0.92709	0.93252
43	0.93252	0.93797	0.94345	0.94896	0.95451	0.96008	0.96569
44°	0.96569	0.97133	0.97700	0.98270	0.98843	0.99420	1.00000

83° 35', or 6° 25'. Table 5 shows the natural tangent of 6° 25' to be approximately 0.112. Now, in Fig. 6-9, lay off *AC*, 10'' in length, perpendicular to *AB*, and from point *C* lay off *CD*, a horizontal line 1.12'' in length. Thus, since angle *CAD* is 6° 25', angle *BAD* will be 83° 35', as required.

Having plotted the corners of the plot by the method of tangents, these points may be checked by measuring the coordinates of the points. This procedure will disclose any great errors.

6-11. **Deed Descriptions.** Frequently it is necessary to plot a parcel of ground from its description in a deed. Deeds of all properties may be found in the office of the Recorder of Deeds located in the county seat or in the municipal offices. In each deed is a description of the property involved. The following deed description relates to the illustrative example for which the computations have been previously given in this book and which is shown in Fig. 8-1.

All That Certain lot or piece of ground with the buildings and improvements thereon erected. *Situate* in Newville, Bucks County, Pennsylvania, beginning at the northeast intersection of Main Street (eighty feet wide) and Chestnut Street (fifty feet wide) and running along the easterly side of said Chestnut Street and measured N 14° 40' W a distance of sixty-two and fifty-four hundredths (62.54) feet to a point at the southwest corner of property now or formerly belonging to William A. Weaver; thence turning and running by the land of said Weaver and measured N 40° 50' E a distance of forty-two and sixty-three hundredths (42.63) feet to a point in the southern boundary of the said Weaver property; thence turning and running by the land of said Weaver and measured S 84° 30' W a distance of eighty-four and thirty-three hundredths (84.33) feet to a point in the western boundary of a property now or formerly belonging to Robert B. Rogers; thence turning and running by the land of said Rogers and measured S 2° 10' E a distance of seventy-five and seven hundredths (75.07) feet to a point in the northerly side of the aforesaid Main Street; thence turning and running along the northerly side of said Main Street and measured N 75° 20' E a distance of one hundred two and seven hundredths (102.07) feet to the first mentioned point and place of beginning.

6-12. **Suggestions.** As a means of understanding thoroughly the methods and procedures discussed in Chapters 4, 5, 6, and 7, nothing is superior to a practical example. Therefore, it is suggested that the student set up three or more stakes in a field and run a traverse through these points. Having made the survey, complete all the necessary computations and plot the results as described in the last example.

CHAPTER 7

COMPUTATION OF AREAS

7-1. Computation of Areas. It is frequently necessary to compute the area of a plot. There are several methods by which this may be accomplished; among them is the *method of coordinates*. In order to apply this method of finding the area of a closed traverse it is necessary that the coordinates of the corners of the plot be established. This procedure is explained in Art. 6-9, and the X and Y coordinates for the corners of our illustrative example are tabulated in Fig. 6-7.

Figure 7-1 shows a four-sided area, $ABCD$, with the X and Y coordinates of points A and D. The coordinates of points B and

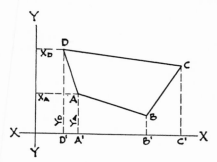

FIG. 7-1.

C are not indicated. By examining this diagram it is seen that the area of $ABCD$ may be computed by finding the area of trapezoid $DD'C'C$ and subtracting from it the sum of trapezoids $DD'A'A$, $AA'B'B$, and $BB'C'C$. The area of a trapezoid is equal to one-half the sum of the parallel sides multiplied by the perpendicular distance between them. By applying this system of computation, an equation for determining the area of the quadrilateral

83

ABCD may be derived. A convenient form of the equation is:

area of $ABCD = \frac{1}{2}[X_A(Y_B - Y_D)$

$$+ X_B(Y_C - Y_A) + X_C(Y_D - Y_B) + X_D(Y_A - Y_C)]$$

Therefore, to find the area of a polygon begin at any corner (point *A*, for example) and proceed in a counterclockwise direction

COMPUTATION OF AREA

$$\text{Area} = \frac{1}{2}\left[X_A(Y_B - Y_E) + X_B(Y_C - Y_A) + X_C(Y_D - Y_B) + X_D(Y_E - Y_C) + X_E(Y_A - Y_D)\right]$$

COR.	COORDINATES		Y-DIFFERENCE		AREAS	
	X	Y	CORNERS	VALUE	+	−
A	15.88	0	B-E	-34.66		550
B	114.63	25.83	C-A	+100.84	11559	
C	111.80	100.84	D-B	+66.92	7482	
D	27.86	92.75	E-C	-40.35		1124
E	0	60.49	A-D	-92.75	0	

$$+19041 \quad -1674$$
$$-1674$$
$$2 \,\overline{)\,17367\,}$$
$$43560 \,\overline{)\,8683.5\,} = \text{Area in sq.ft.}$$
$$0.1993 = \text{Area in acres}$$

log 15.88 = 1.20085
log 34.66 = 1.53983
        2.74068
        = 550.4

log 114.63 = 2.05929
log 100.84 = 2.00363
        4.06292
        = 11559

log 111.80 = 2.04844
log 66.92 = 1.82556
        3.87400
        = 7482

log 27.86 = 1.44498
log 40.35 = 1.60584
        3.05082
        = 1124

Y-DIF.

25.83
− 60.49
− 34.66

100.84
− 0
100.84

92.75
− 25.83
66.92

60.49
−100.84
− 40.35

0
92.75
− 92.75

log 8683.5 = 3.93870
colog 43560 = 5.36091 −10
             9.29961
             = .1993

FIG. 7-2.

(A, B, C, D). Multiply the X coordinate of the beginning point (X_A) by the difference between the following and the preceding Y coordinates $(Y_B - Y_D)$. Similarly, obtain the sum of these products for each point in succession. One-half this sum gives the area of the given polygon. The above equation relates to a four-sided figure, but this method of computation may be applied to a polygon having any number of sides.

The plot of ground in our example, shown in Fig. 8-1, has five corners (five sides); hence, in the above equation, we must include the coordinates of points A, B, C, D, and E. These values are shown in Fig. 6-7 and are again tabulated in Fig. 7-2. The above equation is most readily solved by tabulating the various terms as shown. The plus and minus areas, shown in the last two columns on the right side of the table, are the products of the Y differences and the X coordinates, as required by the formula. For example, the minus 550 shown in the last column of the table is $X_A(Y_B - Y_E)$, the product of $15.88 \times (-34.66)$. The tabulation shows that the area of the polygon $ABCDE$ is 8,683ft^2, or 0.1993 acres.

Another example of computing an area by this method is given in Art. 8-5.

7-2. Areas with Irregular Boundaries. The plot shown in Fig. 8-1 is bounded by straight lines. To compute the area when one or more of the boundaries is an irregular line, a straight line is first run as close as possible to the irregular boundary and offsets are measured from this line. To simplify computations, the offsets should be taken at the same distance apart. Figure 7-3 (a) shows a plot bounded by three straight sides and one irregular side. The area of the quadrangle $ABCD$ may be computed by the method explained in Art. 7-1, and to this area is added the irregular area $EGKH$ and the two triangles DEH and GCK. The area $EGKH$ may be computed by adding together the series of trapezoids.

A more accurate method of finding the area of the irregular area of the plot is to apply *Simpson's one-third rule*, which is based on the assumption that the boundary curve is a series of parabolic curves. By this rule,

$$\text{area} = \frac{d}{3} \left(h_e + 2\Sigma h_{\text{odd}} + 4\Sigma h_{\text{even}} + h_e' \right)$$

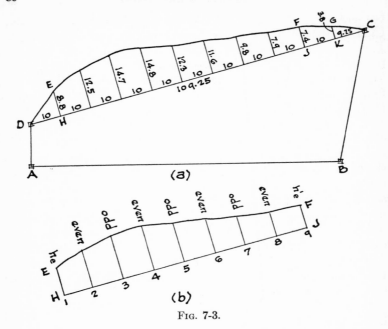

FIG. 7-3.

in which d = distance between offsets. *This distance must be the
same for all offsets.*

h_e = length of the first offset. See Fig. 7-3 (*b*).

$2\Sigma h_{\text{odd}}$ = 2 × sum of the lengths of all the odd offsets.

$4\Sigma h_{\text{even}}$ = 4 × sum of the lengths of all the even offsets.

h_e' = length of the last offset. See Fig. 7-3 (*b*).

*This formula can only be used when there is an even number of
strips.* When there is an odd number of strips the area of the last
strip is computed as a trapezoid, as illustrated in the following
example.

EXAMPLE

EXAMPLE. Compute the area of the irregular portion of the
plot *DEFGC*, shown in Fig. 7-3 (*a*), the dimensions shown being
in feet.

SOLUTION. Simpson's rule will be used. Note that an *even* number of strips will include only the area *EFJH*. This area is again shown in Fig. 7-3 (*b*), in which are shown the even- and odd-numbered offsets as well as the offsets designated as h_e and h_e'. Then

$$\text{area of } EFJH = \tfrac{10}{3}\,[8.8 + (2 \times 36.8) + (4 \times 46.8) + 7.4]$$

$$\text{area of } EFJH = \quad 923.3\text{ft}^2$$

$$\text{area of trapezoid } FGKJ = \frac{7.4 + 3.8}{2} \times 10 = \quad 56.0$$

$$\text{area of triangle } DEH = \frac{10 \times 8.8}{2} = \quad 44.0$$

$$\text{area of triangle } GKC = \frac{9.25 \times 3.8}{2} = \quad 17.6$$

$$\text{area of } DEFGC = 1{,}040.9\text{ft}^2$$

Σh_{odd}	Σh_{even}
14.7	12.5
12.3	14.8
9.8	11.6
	7.9
36.8	46.8

When triangles which are not right triangles occur in the plot, their areas may be found by measuring the length of each of the three sides and using the formula:

$$\text{area} = \sqrt{s(s - a)(s - b)(s - c)}$$

in which a, b, and c are the lengths of the sides and $s = \tfrac{1}{2}(a + b + c)$. See Art. 2–12.

To compute irregular areas, the traverse should be run as close to the irregular boundary line as possible in order to reduce the lengths of the offsets. When the boundary is a fairly smooth

curve the intervals between offsets may be greater than when the
curve is more irregular.   Smaller intervals result in greater accuracy in the computed area.

PROBLEM

7-2-A. Compute the area of land within the boundaries shown in Fig. 7-4.
The dimensions shown are in feet.

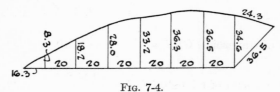

FIG. 7-4.

CHAPTER 8

MISCELLANEOUS SURVEYING PROBLEMS

8-1. Surveying Problems. After the overall dimensions of the buildings to be placed on the plot have been determined, the architect is confronted with the problem of determining the distances of corners of the buildings, roadways, etc., from the boundary lines of the plot. If the intersecting boundary lines of the plot are not right angles, the best method of locating the corners of the building is by the use of coordinates. It is of great assistance to begin by drawing the plot accurately to scale. The survey plot received from the surveyor will show the lengths of the boundary lines and the various angles, but the coordinates of the corners are seldom shown. They should be computed for use in future computations. For the corners of the plot in our illustrative problem the coordinates were computed in Fig. 6-7 and are tabulated in Fig. 7-2.

8-2. Finding Coordinates of the Corners of a Building. Figure 8-1 shows our property plotted with relation to the adjoining streets. The building we are to place on the plot is rectangular in plan, $FGHJ$, and has a length and width of 62' 5" and 28' $3\frac{1}{2}$", respectively. Zoning regulations require that the building must not be closer than 14' 0" to the street line AB or closer than 14' 0" to the party line BC. The architect has decided to place the long axis of the building parallel to the north boundary line CD.

The corner of the building F in relation to the boundary lines AB and BC is shown in Fig. 8-2. Two equal right triangles, FLB and FKB, are formed. Since angles FBK and FBL each equal $\frac{1}{2} \times 102° 30'$, or 51° 15', angles BFK and BFL each equal $180° - (90 + 51° 15')$, or 38° 45'. Then

$$LB = KB = 14 \times \tan 38° 45' = 11.24'$$

$$\log 14 = 1.14613$$

$$\log \tan 38° 45' = 9.90449$$

$$\overline{}$$

$$1.05062$$

$$= 11.24'$$

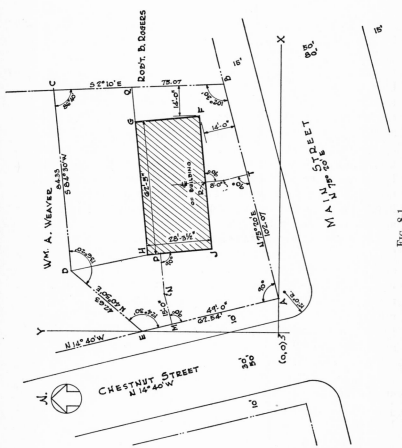

FIG. 8-1.

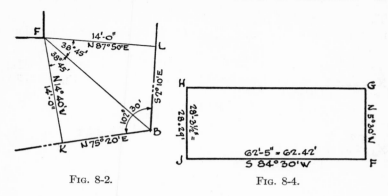

FIG. 8-2.   FIG. 8-4.

The bearings of KB and BL are N 75° 20′ E and S 2° 10′ E, respectively; hence, since angles FKB and FLB are each 90° 0′, the bearings of FK and FL are, respectively, N 14° 40′ W and N 87° 50′ E. We know the coordinates of point B to be $X_B =$ 114.63 and $Y_B = 25.83$ (Fig. 7-2); consequently we can now compute the coordinates of point F by the method explained in Art. 6-3. These computations are shown in Fig. 8-3, and we find $X_F = 100.22$ and $Y_F = 36.53$.

Since GH is parallel to CD, Fig. 8-1, the bearings of lines GH and $FJ = $ S 84° 30′ W. Consequently, the building being rectangular, the bearings of JH and $FG = $ N 5° 30′ W. See Fig. 8-4.

X COORDINATES	DEPARTURES	LATITUDES	Y COORDINATES
$X_B = 114.63$			$Y_B = 25.83$
	log 11.24 = 1.05062	log 11.24 = 1.05062 *	
	log sin 75°20′ = 9.98561	log cos 75°20′ = 9.40346	
	1.03623	0.45408	
-10.87	= 10.87	= 2.845	-2.84
$X_K = 103.76$			$Y_K = 22.99$
	log 14 = 1.14613	log 14 = 1.14613	
	log sin 14°40′ = 9.40346	log cos 14°40′ = 9.98561	
	0.54959	1.13174	
-3.54	= 3.54	= 13.54	$+13.54$
$X_F = 100.22$			$Y_F = 36.53$

* NOTE – Here the actual log as found in the previous computation has been used

FIG. 8-3.

The coordinates of point F having been determined, the coordinates of points G, H, and J can now be found, the computations

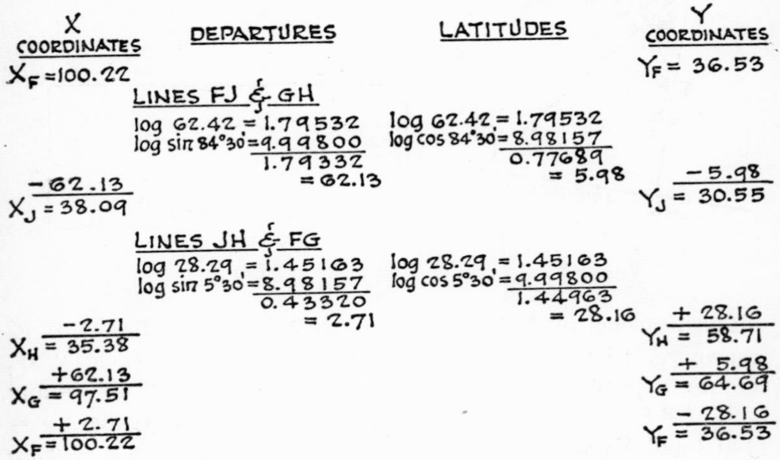

X COORDINATES	DEPARTURES	LATITUDES	Y COORDINATES
$X_F = 100.22$			$Y_F = 36.53$

LINES FJ & GH

log 62.42 = 1.79532   log 62.42 = 1.79532
log sin 84°30' = 9.99800   log cos 84°30' = 8.98157
————————   ————————
1.79332   0.77689
= 62.13   = 5.98

$\dfrac{-62.13}{X_J = 38.09}$ $\dfrac{-5.98}{Y_J = 30.55}$

LINES JH & FG

log 28.29 = 1.45163   log 28.29 = 1.45163
log sin 5°30' = 8.98157   log cos 5°30' = 9.99800
————————   ————————
0.43320   1.44963
= 2.71   = 28.16

$\dfrac{-2.71}{X_H = 35.38}$ $\dfrac{+28.16}{Y_H = 58.71}$

$\dfrac{+62.13}{X_G = 97.51}$ $\dfrac{+5.98}{Y_G = 64.69}$

$\dfrac{+2.71}{X_F = 100.22}$ $\dfrac{-28.16}{Y_F = 36.53}$

Fig. 8-5.

being shown in Fig. 8-5. From these computations we find

$$X_F = 100.22 \quad \text{and} \quad Y_F = 36.53$$

$$X_G = 97.51 \quad \text{and} \quad Y_G = 64.69$$

$$X_H = 35.38 \quad \text{and} \quad Y_H = 58.71$$

$$X_J = 38.09 \quad \text{and} \quad Y_J = 30.55$$

8-3. Length and Bearing of a Missing Line. When the coordinates of the two ends of a line are known, the length and bearing of the line may be found by the methods explained in Art. 6-8.

EXAMPLES

EXAMPLE. It is required to build a fence between points D and H, Fig. 8-1. What is the length of the line DH and what are the angles EDH and HDC?

SOLUTION. The first step is to determine the coordinates of points D and H. This has already been done, and from Fig. 7-2 and Art. 8-2 we see that

$$X_D = 27.86 \quad \text{and} \quad Y_D = 92.75$$

$$X_H = 35.38 \quad \text{and} \quad Y_H = 58.71$$

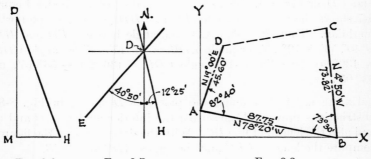

FIG. 8-6.   FIG. 8-7.   FIG. 8-8.

Referring to the right triangle shown in Fig. 8-6, it is seen that

distance $MH = X_H - X_D = 35.38 - 27.86 = 7.52$

distance $DM = Y_D - Y_H = 92.75 - 58.71 = 34.04$

Then   distance $DH = \sqrt{7.52^2 + 34.04^2} = \sqrt{56.55 + 1{,}158.72}$

distance $DH = \sqrt{1{,}215.27} = 34.86'$

$$\log 1{,}215 = 3.08458$$
$$\div 2$$
$$\overline{}$$
$$1.54229$$
$$= 34.86$$

tan bearing of DH (Fig. 8-6) $= \dfrac{MH}{DM} = \dfrac{7.52}{34.04}$ (see Art. 6-8)

bearing of $DH = $ S 12° 25′ E

$$\log 7.52 = 0.87622$$
$$\text{colog } 34.04 = 8.46801$$
$$\overline{}$$
$$9.34423$$
$$= 12° 25'$$

Note that the bearing is taken to the nearest 05', to be consistent with the accuracy of the survey.

Now that the bearings of the three intersecting lines *ED*, *HD* and *CD* are known, the angles are readily established. Figure 8-7 shows the angles that lines *ED* and *HD* make with the north meridian. Thus, it is seen that the angle between *ED* and *HD* is 40° 50' + 12° 25', or angle *EDH* = 53° 15'. Since angle *CDE* is 136° 20' (see Fig. 8-1), angle *CDH* = 136° 20' − 53° 15', or 83° 05'.

EXAMPLE. In surveying the plot *ABCD* shown in Fig. 8-8, obstructions prevented measuring the interior angles at *D* and *C* and also the length of line *DC*. With the data shown in the figure, compute the length of *DC* and the angles *ADC* and *DCB*.

SOLUTION. STEP 1. First determine the coordinates of points *D* and *C*. Since the bearings and lengths of *DA*, *AB*, and *BC*

X COORDINATES	DEPARTURES	LATITUDES	Y COORDINATES
$X_A = 0$			$Y_A = 17.74$
	LINE AB		
	log 87.75 = 1.94325	log 87.75 = 1.94325	
	log sin 78°20' = 9.99093	log cos 78°20' = 9.30582	
	1.93418	1.24907	
+85.94	= 85.94	= 17.74	−17.74
$X_B = 85.94$			$Y_B = 0$
	LINE BC		
	log 73.82 = 1.86817	log 73.82 = 1.86817	
	log sin 4°50' = 8.92561	log cos 4°50' = 9.99845	
	0.79378	1.86662	
−6.22	= 6.22	= 73.56	+73.56
$X_C = 79.72$			$Y_C = 73.56$
	LINE DA		
$X_A = 0$	log 45.60 = 1.65896	log 45.60 = 1.65896	$Y_A = 17.74$
	log sin 19°00' = 9.51264	log cos 19°00' = 9.97567	
	1.17160	1.63463	
+14.85	= 14.85	= 43.12	43.12
$X_D = 14.85$			$Y_D = 60.86$

FIG. 8-9.

have been established, the coordinates of points *D* and *C* may be found by the methods previously explained. The necessary computations are shown in Fig. 8-9. Note that the coordinates of *D* and *C* are

$$X_D = 14.85 \quad \text{and} \quad Y_D = 60.86$$
$$X_C = 79.72 \quad \text{and} \quad Y_C = 73.56$$

STEP 2. Compute the length of DC. Now that the coordinates of D and C have been found, the right triangle, shown in Fig. 8-10, shows that

$$\text{departure } DC' = 79.72 - 14.85 = 64.87$$
$$\text{latitude } CC' = 73.56 - 60.86 = 12.70$$

Hence, length of $DC = \sqrt{64.87^2 + 12.70^2} = \sqrt{4{,}369} = 66.10'$

STEP 3. Compute the bearing of DC.

$$\tan \text{ bearing of } DC = \frac{\text{departure}}{\text{latitude}} = \frac{64.87}{12.70} \qquad \text{(see Art. 6-8)}$$

bearing of $DC = \text{N } 78° \, 55' \text{ E}$

$$\log 64.87 = 1.81204$$
$$\text{colog } 12.70 = 8.89620$$
$$\overline{\qquad\qquad 0.70824}$$
$$= 78° \, 55'$$

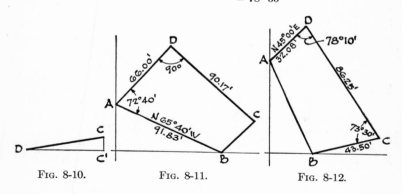

FIG. 8-10.   FIG. 8-11.   FIG. 8-12.

STEP 4. Compute angles ADC and DCB. We now know the bearings of lines AD, DC, and CB and, by the method explained in Art. 5-3, we find angle $ADC = 120° \, 05'$ and angle $DCB = 83° \, 45'$. To check these angles we can use the rule given in Art.

5-6. The sum of the interior angles $= 180(n - 2) = 180(4 - 2)$ $= 360°$. Thus $73° 30' + 82° 40' + 120° 05' + 83° 45' = 360° 0'$.

PROBLEMS

8-3-A-B. In the surveys shown in Fig. 8-11 and 8-12 compute the magnitudes of the missing interior angles and the bearings and lengths of the missing sides.

8-4. **Missing Line Problems.** Many problems relating to a missing line may be solved by first finding the coordinates of the ends of the line.

EXAMPLES

EXAMPLE. In Fig. 8-1 line MNP represents the center line of a pathway. Point M is $49' 0''$ from point A, and MN is $15' 0''$ in length. Determine the length of NP and the point at which the center of the pathway intersects the building, the length HP.

SOLUTION. STEP 1. Determine the coordinates of point H. These coordinates were established in Art. 8-2. We found that

$$X_H = 35.38 \quad \text{and} \quad Y_H = 58.71$$

STEP 2. Determine the coordinates of point N. We know the bearing of line AE to be N 14° 40' W, and, since MN makes an angle of 90° 0' with AE, the bearing of $MN = 90° 0' - 14° 40' =$ N 75° 20' E. In Fig. 7-2 we find that $X_A = 15.88$ and $Y_A = 0$.

X COORDINATES	DEPARTURES	LATITUDES	Y COORDINATES
$X_A = 15.88$			$Y_A = 0$

LINE AM

log 49 $= 1.69020$		log 49 $= 1.69020$
log sin 14°40' $= 9.40346$		log cos 14°40' $= 9.98561$
$\overline{1.09366}$		$\overline{1.67581}$
$= 12.41$		$= 47.40$

$\dfrac{-12.41}{X_M = 3.47}$ $\qquad\qquad\qquad\qquad\qquad \dfrac{+47.40}{Y_M = 47.40}$

LINE MN

log 15 $= 1.17609$		log 15 $= 1.17609$
log sin 75°20' $= 9.98561$		log cos 75°20' $= 9.40346$
$\overline{1.16170}$		$\overline{0.57955}$
$= 14.51$		$= 3.80$

$\dfrac{+14.51}{X_N = 17.98}$ $\qquad\qquad\qquad\qquad\qquad \dfrac{+3.80}{Y_N = 51.20}$

FIG. 8-13.

With these data we can now compute the coordinates of point N. This is shown in Fig. 8-13, and we find that $X_N = 17.98$ and $Y_N = 51.20$.

STEP 3. Find the length and bearing of NH.

departure of $NH = 35.38 - 17.98 = 17.40$

latitude of $NH = 58.71 - 51.20 = 7.51$

length of $NH = \sqrt{17.40^2 + 7.51^2} = \sqrt{359.2} = 18.95'$

$$\log 359.2 = 2.55534$$
$$\div 2$$
$$\overline{1.27767}$$

$$= 18.95$$

tan bearing $NH = \dfrac{\text{departure}}{\text{latitude}} = \dfrac{17.40}{7.51}$ (see Art. 6-8)

bearing of $NH =$ N 66° 40′ E

$$\log 17.40 = 1.24055$$
$$\text{colog } 7.51 = 9.12436$$
$$\overline{0.36491}$$

$$= 66° 40'$$

STEP 4. Find the interior angle at H shown in the right triangle, Fig. 8-14. The lines JH and FG shown in Fig. 8-4 are parallel;

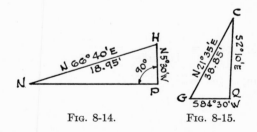

FIG. 8-14.   FIG. 8-15.

hence the bearing of JH and PH (Fig. 8-14) is N 5° 30′ W. Therefore, angle $NHP = 66° 40' + 5° 30' = 72° 10'$.

STEP 5. Find HP and NP by solving the right triangle NHP, Fig. 8-14.

$$HP = 18.95 \times \cos 72°\ 10' = 5.80' = 5'\ 9\tfrac{5}{8}''$$

$$NP = 18.95 \times \sin 72°\ 10' = 18.04' = 18'\ 0\tfrac{1}{2}''$$

$\log 18.95 = 1.27761$	$\log 18.95 = 1.27761$
$\log \cos 72°\ 10' = 9.48607$	$\log \sin 72°\ 10' = 9.97861$
0.76368	1.25622
$= 5.80$	$= 18.04$

EXAMPLE. It is desired to build a fence from G to Q shown on Fig. 8-1. The fence is a continuation of the building line HG. Compute the length of the fence GQ and the distance CQ.

SOLUTION. STEP 1. Compute the length and bearing of the line GC.

$$X_C = 111.80 \quad \text{and} \quad Y_C = 100.84 \quad \text{(from Fig. 7-2)}$$

$$X_G = 97.51 \quad \text{and} \quad Y_G = 64.69 \quad \text{(from Art. 8-2)}$$

departure   latitude

of GC = 14.29   of GC = 36.15

length of $GC = \sqrt{14.29^2 + 36.15^2} = \sqrt{1{,}509} = 38.85'$

$$\log 1{,}509 = 3.17869$$
$$\div\ 2$$
$$1.58935$$
$$= 38.85$$

tan bearing $GC = \dfrac{\text{departure}}{\text{latitude}} = \dfrac{14.29}{36.15}$ (see Art. 6-8)

bearing of $GC = $ N $21°\ 35'$ E

$$\log 14.29 = 1.15503$$
$$\text{colog } 36.15 = 8.44189$$
$$9.59692$$
$$= 21°\ 35'$$

STEP 2.   Compute angles at G, C, and Q shown in Fig. 8-15. Since the bearings of the three sides of the triangle are known,

$$\text{angle } CGQ = 84° \, 30' - 21° \, 35' \qquad = \quad 62° \, 55'$$

$$\text{angle } QCG = 21° \, 35' + 2° \, 10' \qquad = \quad 23° \, 45'$$

$$\text{angle } GQC = 180° - 2° \, 10' - 84° \, 30' = \quad 93° \, 20'$$

$$\overline{\qquad\qquad\qquad\qquad\qquad}$$

$$180° \, 00'$$

STEP 3.   Compute the lengths of lines CQ and GQ. These lengths are most readily found by the application of the sine law given in Art. 2-11. This law states that in any triangle the sides are proportional to the sines of the opposite angles. The computations for these lengths are shown in Fig. 8-16, and it is found that $CQ = 34.65'$ and $GQ = 15.67'$.

$$\frac{38.85}{\sin 93°20'} = \frac{CQ}{\sin 62°55'} = \frac{CQ}{\sin 23°45'}$$

$$\text{Length } CQ = \frac{38.85 \times \sin 62°55'}{\sin 93°20'} = 34.65'$$
$$\text{or } 34'\text{-}7\tfrac{3}{4}''$$

Note: $\sin 93°20' = -\sin(180° - 93°20') = -\sin 86°40'$

$$\text{Length } GQ = \frac{38.85 \times \sin 23°45'}{\sin 93°20'} = 15.67'$$
$$\text{or } 15'\text{-}8''$$

log 38.85 = 1.58939
log sin 62°55' = 9.94956
colog sin 86°40' = 0.00074
1.53969
= 34.65

log 38.85 = 1.58939
log sin 23°45' = 9.60503
colog sin 86°40' = 0.00074
1.19516
= 15.67

FIG. 8-16.

PROBLEM

8-4-A. The line RST in Fig. 8-1 represents the center line of an entrance pathway.   If the distance RS is 8' 0", compute the lengths of ST and TB.

8-5. Areas of Plots.   It is frequently necessary to compute the magnitudes of certain areas within the boundaries of a plot. These areas may often be computed by first dividing them into simple figures and then finding their sum. When, however, the coordinates of the various points of the survey have been previously found it may be convenient to use the system of computation given in Art. 7-1.

EXAMPLE

EXAMPLE. It is required to cover the area $HQCD$, Fig. 8-1, with bituminous paving. How many square yards of paving are required?

SOLUTION. The coordinates of points C and D are found in Fig. 6-7, and the coordinates of point H are given in Art. 8-2. It remains, therefore, to compute the coordinates of point Q. From Art. 8-4 it was found that GQ has a length of 15.67′; its bearing is S 84° 30′ W. The computations are shown in Fig. 8-17; $X_Q = 113.11$ and $Y_Q = 66.19$.

X COORDINATES	DEPARTURES	LATITUDES	Y COORDINATES
$X_G = 97.51$			$Y_G = 64.69$
	log 15.67 = 1.19507	log 15.67 = 1.19507	
	log sin 84°30′ = 9.99800	log cos 84°30′ = 8.98157	
	1.19307	0.17664	
+15.60	= 15.60	= 1.50	+1.50
$X_Q = 113.11$			$Y_Q = 66.19$

FIG. 8-17.

With these data we can use the formula given in Art. 7-1. The simplest method of using this formula is to tabulate the terms;

COMPUTATION OF AREAS

COR.	COORDINATES		Y-DIFFERENCE		AREAS	
	X	Y	CORNER	VALUE	+	−
H	35.38	58.71	Q−D	−26.56		940
Q	113.11	66.19	C−H	+42.13	4765	
C	111.80	100.84	D−Q	+26.56	2969	
D	27.86	92.75	H−C	−42.13		1174

$$+7\,734 \quad -2\,114$$
$$-2\,114$$
$$2\,|\,5\,620$$
$$9\,|\,2\,810 = \text{Area in sq. ft.}$$
$$312.2 = \text{Area in sq. yds.}$$

FIG. 8-18.

this is shown in Fig. 8-18, and we find that area $HQCD$ contains 312.2 square yards.

PROBLEM

8-5-A. Compute the area of $HQCD$, Fig. 8-1, by considering it to be a trapezoid. Note that GQ was computed to be 15.67′. (Suggestion: Compute the perpendicular distance between the parallel lines DC and HQ.)

CHAPTER 9

CIRCULAR CURVES

9-1. Circular Curves. The curves commonly employed in building operations are *horizontal circular curves*, arcs of circles. Figure 9-1 shows two non-parallel lines, *A-P.C.* and *B-P.T.*, that intersect at point *V*, the *vertex*. A circular curve, whose radius is *R* and whose center is point *O*, connects the two straight lines. Point *P.C.*, the *point of curvature*, is the point at which the curve is tangent to the line *AV* and at which the circular curve begins. Point *P.T.*, the *point of tangency*, is the point at which the curve is tangent to the line *BV* and at which the straight line *P.T.-B* begins. The straight lines joining the curve are called *tangents*. Since the curve is tangent to lines *AV* and *BV* at points *P.C.*

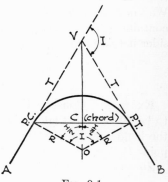

FIG. 9-1.

and *P.T.*, the angles *A-P.C.-O* and *B-P.T.-O* are right angles. The angle between the two radii at point *O* is angle *I*, the *included angle*. *I* is also always equal to the external angle at point *V*. The distances from *V* to *P.C.* and from *V* to *P.T.* are equal; this distance is *T*, the *tangent distance*.

If we know *R*, the radius of the arc, and also *I*, the included angle, the tangent distance *T* may be computed by the formula

$$T = R \tan \frac{I}{2}$$

The straight line connecting points *P.C.* and *P.T.* is *C*, the *chord*. Its length may be computed by the formula

$$C = 2R \sin \frac{I}{2}$$

9-2. Length of Curves. To designate a specific curve, three elements should be given. They are

(*a*) The radius of the curve *R*.
(*b*) The included angle *I*.
(*c*) The length of the arc.

Whereas any two of these elements identify the curve, all three should be computed and shown on the drawings.

When the radius of the curve and the included angle are known, the length of the arc may be computed by two different methods. We know the *circumference* of a circle to be $2\pi R$ and that a circle contains 360°. Thus, to find the length of an arc, we simply consider it to be a certain part of a circumference.

EXAMPLES

EXAMPLE. Consider the radius of the curve, shown in Fig. 9-1, to be 40' 0'' and that *I*, the included angle, is 85° 20'. Compute the length of the arc, the length of the curve from point *P.C.* to point *P.T.*

SOLUTION. Since 85° 20' may be written 85.33°,

$$\frac{85.33}{360} \times 2\pi R = \frac{85.33}{360} \times 2 \times 3.1416 \times 40$$

$$= 59.57', \text{ the length of the curve}$$

A simple method of determining the length of a curve is to make use of Table 6, a table giving the lengths of circular arcs. In using this table, note that the given lengths of arcs are for arcs whose radius is 1. To use the table, add together the lengths (in the table) opposite the degrees, minutes, and seconds corresponding to the included angle, and multiply their sum by the

TABLE 6. LENGTHS OF CIRCULAR ARCS. RADIUS = 1

Sec	Length	Min	Length.	Deg	Length	Deg	Length
1	0.0000048	1	0.0002909	1	0.0174533	61	1.0646508
2	0.0000097	2	0.0005818	2	0.0349066	62	1.0821041
3	0.0000145	3	0.0008727	3	0.0523599	63	1.0995574
4	0.0000194	4	0.0011636	4	0.0698132	64	1.1170107
5	0.0000242	5	0.0014544	5	0.0872665	65	1.1344640
6	0.0000291	6	0.0017453	6	0.1047198	66	1.1519173
7	0.0000339	7	0.0020362	7	0.1221730	67	1.1693706
8	0.0000388	8	0.0023271	8	0.1396263	68	1.1868239
9	0.0000436	9	0.0026180	9	0.1570796	69	1.2042772
10	0.0000485	10	0.0029089	10	0.1745329	70	1.2217305
11	0.0000533	11	0.0031998	11	0.1919862	71	1.2391838
12	0.0000582	12	0.0034907	12	0.2094395	72	1.2566371
13	0.0000630	13	0.0037815	13	0.2268928	73	1.2740904
14	0.0000679	14	0.0040724	14	0.2443461	74	1.2915436
15	0.0000727	15	0.0043633	15	0.2617994	75	1.3089969
16	0.0000776	16	0.0046542	16	0.2792527	76	1.3264502
17	0.0000824	17	0.0049451	17	0.2967060	77	1.3439035
18	0.0000873	18	0.0052360	18	0.3141593	78	1.3613568
19	0.0000921	19	0.0055269	19	0.3316126	79	1.3788101
20	0.0000970	20	0.0058178	20	0.3490659	80	1.3962634
21	0.0001018	21	0.0061087	21	0.3665191	81	1.4137167
22	0.0001067	22	0.0063995	22	0.3839724	82	1.4311700
23	0.0001115	23	0.0066904	23	0.4014257	83	1.4486233
24	0.0001164	24	0.0069813	24	0.4188790	84	1.4660766
25	0.0001212	25	0.0072722	25	0.4363323	85	1.4835299
26	0.0001261	26	0.0075631	26	0.4537856	86	1.5009832
27	0.0001309	27	0.0078540	27	0.4712389	87	1.5184364
28	0.0001357	28	0.0081449	28	0.4886922	88	1.5358897
29	0.0001406	29	0.0084358	29	0.5061455	89	1.5533430
30	0.0001454	30	0.0087266	30	0.5235988	90	1.5707963
31	0.0001503	31	0.0090175	31	0.5410521	91	1.5882496
32	0.0001551	32	0.0093084	32	0.5585054	92	1.6057029
33	0.0001600	33	0.0095993	23	0.5759587	93	1.6231562
34	0.0001648	34	0.0098902	34	0.5934119	94	1.6406095
35	0.0001697	35	0.0101811	35	0.6108652	95	1.6580628
36	0.0001745	36	0.0104720	36	0.6283185	96	1.6755161
37	0.0001794	37	0.0107629	37	0.6457718	97	1.6929694
38	0.0001842	38	0.0110538	38	0.6632251	98	1.7104227
39	0.0001891	39	0.0113446	39	0.6806784	99	1.7278760
40	0.0001939	40	0.0116355	40	0.6981317	100	1.7453293
41	0.0001988	41	0.0119264	41	0.7155850	101	1.7627825
42	0.0002036	42	0.0122173	42	0.7330383	102	1.7802358
43	0.0002085	43	0.0125082	43	0.7504916	103	1.7976891
44	0.0002133	44	0.0127991	44	0.7679449	104	1.8151424
45	0.0002182	45	0.0130900	45	0.7853982	105	1.8325957
46	0.0002230	46	0.0133809	46	0.8028515	106	1.8500490
47	0.0002279	47	0.0136717	47	0.8203047	107	1.8675023
48	0.0002327	48	0.0139626	48	0.8377580	108	1.8849556
49	0.0002376	49	0.0142535	49	0.8552113	109	1.9024089
50	0.0002424	50	0.0145444	50	0.8726646	110	1.9198622
51	0.0002473	51	0.0148353	51	0.8901179	111	1.9373155
52	0.0002521	52	0.0151262	52	0.9075712	112	1.9547688
53	0.0002570	53	0.0154171	53	0.9250245	113	1.9722221
54	0.0002618	54	0.0157080	54	0.9424778	114	1.9896753
55	0.0002666	55	0.0159989	55	0.9599311	115	2.0071286
56	0.0002715	56	0.0162897	56	0.9773844	116	2.0245819
57	0.0002763	57	0.0165806	57	0.9948377	117	2.0420352
58	0.0002812	58	0.0168715	58	1.0122910	118	2.0594885
59	0.0002860	59	0.0171624	59	1.0297443	119	2.0769418
60	0.0002909	60	0.0174533	60	1.0471976	120	2.0943951

radius of the circle. As an example, compute the length of the
arc of a circle the radius of which is 20′ 0″, the included angle
being 52° 18′ 40″. Then, referring to Table 6,

length for 52° =   0.9075712

length for 18′ =   0.0052360

length for 40″ =   0.0001939
                   ──────────
                   0.9130011
                 × 20
                   ──────────
18.2600220 = 18.26′   the length of the arc

EXAMPLE. Compute, by the use of Table 6, the length of the
arc whose radius is 40′ 0″ and whose included angle is 85° 20′.

SOLUTION. Referring to Table 6,

length for 85° =   1.4835299

length for 20′ =   0.0058178
                   ──────────
                   1.4893477
                 × 40
                   ──────────
59.5739080 = 59.57′   the length of the arc

Note that this is the same length found in the first example.

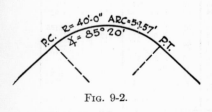

FIG. 9-2.

Now that length of curve
has been found, all three elements of the curve are known
and this information is shown
on the drawing as indicated in
Fig. 9-2. If the vertex and
points of curvature or tangency are tied in by dimen-
sions with other established points on the plot, no other information is needed for staking out the curve.

EXAMPLE. Two intersecting lines and their bearings are shown in Fig. 9-3. A circular curve, whose radius is 50′ 0″, is to be constructed tangent to these lines. Compute the data required to dimension the curve completely.

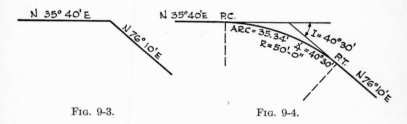

FIG. 9-3.   FIG. 9-4.

SOLUTION. The first step in the solution is to find I, the included angle. Since we know that I is also the magnitude of the external angle at the intersection of the two lines (see Fig. 9-1 and 9-4) and as the bearings of the lines are given, angle I can be computed by means of the principles explained in Art. 5-4. Thus, by making a drawing showing the bearings of the lines, we see that $76° 10′ − 35° 40′ = 40° 30′$, the angle I. Now that I is known, we may use Table 6 to find the length of the curve. Therefore

length for 40° =   0.6981317

length for 30° =   0.0087266
                   ─────────────
                   0.7068583
                   × 50
                   ─────────────

35.3429150 = 35.34′   the length of the curve

The complete curve dimensions are shown in Fig. 9-4.

This example is the type of problem commonly encountered. A variant consists in having as data the length of the curve and the included angle, the radius being unknown.

EXAMPLE. The included angle of a circular curve tangent to two straight lines is $156° 35′ 20″$, and the length of the curve is 62.28′. Compute the radius.

SOLUTION. By the use of Table 6 we find the length of the arc for an angle of 156° 35′ 20″ *if the radius = 1′ 0″*. Then

length for 100° = 1.7453293

length for 56° = 0.9773844

length for 35′ = 0.0101811

length for 20″ = 0.0000970
_____

2.7329918   the length of the curve
for a radius of 1′ 0″

Then, since the length of the arc is 62.28′, $R = \dfrac{62.28}{2.733} = 22.79'$,

the radius of the circular curve.

PROBLEMS

9-2-A-B-C-D. Each diagram shown in Fig. 9-5 gives the bearings of two intersecting lines and the radius of the tangent circular curves. Compute the lengths of the arcs.

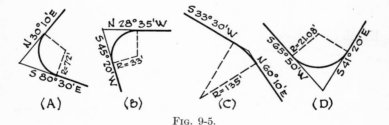

FIG. 9-5.

In the following problems both the lengths of the arcs and the included angles are given. Compute the radii.

PROBLEM	LENGTH OF ARC	INCLUDED ANGLE
9-2-E.	93.46′	130° 33′
9-2-F.	32′ 2½″	47° 30′ 15″
9-2-G.	86.33′	126° 23′
9-2-H.	22′ 7⅝″	33° 16′ 20″

9-3. Laying Out Circular Curves. In laying out curves in the field, stakes are driven on the curve at a sufficient number of points to mark its exact location. Curves having comparatively small radii require a greater number of stakes than curves in which the radii are large. The method commonly employed in computing the location of the points on the curve is the *deflection angle method.*

9-4. Deflection Angles. With respect to curves, a *deflection angle* is the angle that a chord makes with a tangent. Consider any

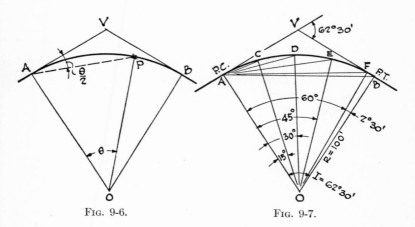

FIG. 9-6.   FIG. 9-7.

point, such as P, on curve AB shown in Fig. 9-6. The deflection angle of point P is angle VAP. *The deflection angle is always one-half the angle subtended by the chord.* In Fig. 9-6, angle $VAP = \frac{1}{2}$ angle POA.

To stake out a curve, the deflection angles and chord lengths are computed for a number of points on the curve. Then $P.C.$, the point of curvature, $P.T.$, the point of tangency and V, the vertex, are all located on the plot and marked with stakes. The transit is set up over the point of curvature (or point of tangency) and sighted on V, the vertex. The deflection angles are then turned off, and the chord distances are measured along these lines, thus determining the positions of the stakes on the curve. Generally a number of equal arcs are laid off with a shorter arc remaining at the end of the curve, as shown in Fig. 9-7. The procedure of computing the deflection angles and chord lengths is illustrated in the following example.

EXAMPLE

EXAMPLE. A curve having a radius of $100' \, 0''$ is tangent to two intersecting lines for which the external angle at their point of intersection, the included angle, is $62° \, 30'$, as shown in Fig. 9-7. Compute the location of four points on the curve.

SOLUTION. Since the external angle at V is $62° \, 30'$, this is also I, the included angle at point O, the center of the circle of which AB is the arc. A line is now drawn from point O to point A (this point is also $P.C.$) and from this line we construct angles of $15°$, $30°$, $45°$, and $60°$, the lines of these angles intersecting the curve at points C, D, E, and F, respectively.

In Art. 9-1 we found that the length of a chord may be found by the formula

$$C \text{ (the chord)} = 2R \sin \frac{I}{2}$$

This formula, coupled with the fact that the deflection angle is one-half the angle subtended by the chord, enables us to compute the various chord lengths.

In Fig. 9-7 chord AC subtends an angle of $15° \, 0'$ and therefore deflection angle $VAC =$ one-half $15° \, 0'$, or $7° \, 30'$. Now, since $C = 2R \sin \dfrac{I}{2}$,

$AC = 2 \times 100 \times \sin 7° \, 30' = 26.11'$ the length of chord AC

Similarly,

$$AD = 2 \times 100 \times \sin 15° \, 0' \ = \ \ 51.76'$$

$$AE = 2 \times 100 \times \sin 22° \, 30' = \ \ 76.54'$$

$$AF = 2 \times 100 \times \sin 30° \, 0' \ = \ 100.0' \quad (\sin 30° = 0.5)$$

To check the remaining chord FB,

$$FB = 2 \times 100 \sin 1° \, 15' = 4.36'$$

The chord lengths given above were computed by the use of logarithms as follows:

$$\begin{array}{ll} \log 200 & = 2.30103 \\ \log \sin 7° \, 30' & = 9.11570 \\ \hline & 1.41673 \\ & = 26.11 \end{array}$$

$$\begin{array}{ll} \log 200 & = 2.30103 \\ \log \sin 15° & = 9.41300 \\ \hline & 1.71403 \\ & = 51.76 \end{array}$$

$$\begin{array}{ll} \log 200 & = 2.30103 \\ \log \sin 22° \, 30' & = 9.58284 \\ \hline & 1.88387 \\ & = 76.54 \end{array}$$

$$\begin{array}{ll} \log 200 & = 2.30103 \\ \log \sin 1° \, 15' & = 8.33875 \\ \hline & 0.63978 \\ & = 4.36 \end{array}$$

Now that the chord lengths have been determined, the instrument is set up over point $P.C.$, sighted on point V, and the angle VAC, $7° \, 30'$, is laid off. On this sighted line the distance AC, $26.11'$, is measured, thus establishing point C on the curve. The remaining points, D, E, and F are located in a similar manner.

In laying out points on the curve, obstructions may obscure the lines of sight. For example, in Fig. 9-7 a tree or other obstacle might prevent points E and F from being visible from $P.C.$ For such a condition points C and D could be located as previously described and the points E and F located from $P.T.$

It is sometimes convenient to divide the curve into a number of equal parts. This procedure is followed when the resulting deflection angles are not beyond the limitations of the surveying instrument.

PROBLEM

9-4-A. For the curve shown in Fig. 9-7, four stakes are to be set so that arc AB is divided into five equal parts. Compute the various deflection angles and the chord distances.

9-5. Laying Out a Given Arc Distance. In order to locate full stations or other exact points on a curve, it is sometimes necessary to lay out a given arc distance from point $P.C.$ Suppose, for example, the *arc* distance $P.C.\text{-}P$, in Fig. 9-8, is given and we are required to compute the *chord* distance $P.C.\text{-}P$. To accomplish this, we first find the subtended angle of the arc $P.C.\text{-}P$; this is angle POA. We know that the deflection angle VAP is one-half

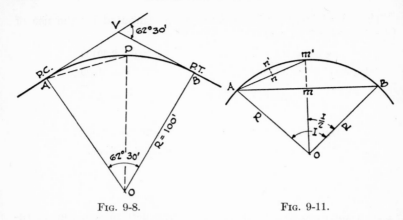

FIG. 9-8.   FIG. 9-11.

angle *POA*, and this information enables us to compute the chord
distance *P.C.-P.*

EXAMPLE

EXAMPLE. In the curve shown in Fig. 9-8, the *arc* distance is
60′ 0″ from point *P.C.* to point *P* and the radius of the curve is
100′ 0″. Compute the deflection angle of point *P* and also the
chord distance, *P.C.-P.*

SOLUTION. The first step is to find the angle subtended by the
arc of 60′ 0″.

The arc for a 100′ 0″ radius = 60′ 0″; hence, the arc for a 1′ 0″
radius = $\frac{60}{100}$ = 0.6000000′ 0″.

Then, from Table 6, the closest arc length to 0.6 is 0.5934119;
it subtends an angle of 34°, the difference in arc length being
0.0065881. Proceeding in this manner,

		0.6000000
from Table 6,	34° =	0.5934119
	difference =	0.0065881
from Table 6,	22′ =	0.0063995
	difference =	0.0001886
from Table 6,	38″ =	0.0001842
		0.0000044

we find that angle POA is 34° 22′ 38″. Therefore, the deflection angle is one-half 34° 22′ 38″, or 17° 11′ 19″, say 17° 11′.

$$\text{chord} = 2 \times R \times \sin \frac{I}{2}$$

$$= 2 \times 100 \times \sin 17° 11′ = 59.09′ \quad \text{the chord distance}$$

Now that both the deflection angle and chord distance have been computed, the instrument is set up over point $P.C.$ and sighted on V. The deflection angle VAP is turned off as close to 17° 11′ as the instrument will permit, and point P is staked out at 59.09′ from point $P.C.$

In the above computations the arc length for a 1′ 0″ radius was taken to be 0.6000000′. This distance should be computed to 7 decimal places to correspond to the arc lengths given in Table 6. Many surveying instruments do not permit laying off angles of degrees, minutes, and seconds. But the reason for carrying the computations to a high degree of accuracy is that very slight variations in laying off the deflection angle result in great discrepancies in locating the desired point on the curve.

PROBLEMS

9-5-A. On the curve shown in Fig. 9-9, it is desired to locate a point at an arc distance of 53.27′ from point $P.C.$ Compute the deflection angle and the chord distance.

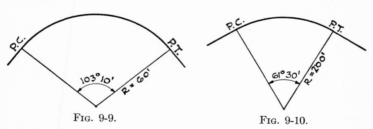

Fig. 9-9.   Fig. 9-10.

9-5-B. On the curve shown in Fig. 9-10, it is desired to locate a point at an arc distance of 153.0′ from point $P.C.$ Compute the deflection angle and the chord distance.

9-6. Middle Ordinates of Circular Curves.

When two points on a circular curve have been located and the chord distance between them is known, other points between them on the curve are found without difficulty. In Fig. 9-11, A and B are two points on a circular curve the radius of which is R. The chord distance may be computed as explained in Art. 9-1 or found by direct measurement in the field. Point m is located at the midpoint of chord AB, and line mm' is set off at right angles to AB; mm' is called the *middle ordinate* and its length is readily computed.

In Art. 9-1 we found the formula for computing the chord distance to be: chord distance $= 2 \times R \times \sin \dfrac{I}{2}$; therefore,

$$\sin \frac{I}{2} = \frac{\frac{1}{2} \text{ chord distance}}{R} \tag{1}$$

The length of the middle ordinate mm' may be found by the formula

$$mm' = R - R \cos \frac{I}{2} \tag{2}$$

EXAMPLES

EXAMPLE. In Fig. 9-11 the radius of the circular curve is $120'\ 0''$ and the chord distance AB is $173.2'$. Compute the length of mm', the middle ordinate.

Solution. The first step is to find angle $\dfrac{I}{2}$. From equation (1),

$$\sin \frac{I}{2} = \frac{86.6}{120}$$

therefore, $\qquad \dfrac{I}{2} = 46°\ 12'$

$$\log 86.6 = 1.93752$$
$$\text{colog } 120 = 7.92082$$
$$\overline{\phantom{\text{colog } 120 = }9.85834}$$

$$= 46°\ 12' \text{ (Table 2)}$$

To find the length of mm', we use equation (2). Then

$$mm' = 120 - (120 \times \cos 46° 12')$$

and $mm' = 120 - 83.06 = 36.94'$ the length of the middle
ordinate mm'

$$\log 120 = 2.07918$$
$$\log \cos 46° 12' = 9.84020$$
$$\overline{}$$
$$1.91938$$
$$= 83.06$$

In order to lay out the curve with greater accuracy, the process
may be repeated and successive middle ordinates, such as nn' in
Fig. 9-11, may be found. Usually, the subtended angle I will not
be as large as that given in the previous example. *When the subtended angle I is less than* 20°, the following rule for finding the
length of the middle ordinate gives lengths that are approximately
correct and sufficiently accurate for most conditions:

$$\text{middle ordinate} = \frac{(\text{chord distance})^2}{8 \times R} \qquad (3)$$

EXAMPLE. In the circular curve shown in Fig. 9-7, compute
the length of the middle ordinate of chord AC by use of equation
(3).

SOLUTION. From the computations shown in Art. 9-4, the
length of chord AC was found to be 26.11′; its subtended angle is
15° and the radius of the curve is 100′ 0″. From equation (3),

$$\text{middle ordinate} = \frac{(26.11)^2}{8 \times 100} = 0.852', \text{ the approximate length}$$

This figure was computed to the third decimal place in order to
compare it with 0.856′, the exact length of the middle ordinate
found by the use of equation (2).

When successive middle ordinates such as nn' in Fig. 9-11 are
sought, the following rule gives results that are approximately
correct;

$$nn' = \frac{\text{middle ordinate distance } (mm')}{4} \qquad (4)$$

EXAMPLE. In the last example the length of the middle ordinate for chord AC was computed to be 0.85′. By the rule expressed by equation (4), compute the length of a sub-middle ordinate. Then

$$nn' = \frac{0.85}{4} = 0.21'$$

The approximations that result from the use of equations (3) and (4) are of convenience to the draftsman when laying out curves of large radius. The chord distances may be scaled from the drawing and the computations performed by use of the slide rule.

PROBLEMS

9-6-A-B. For the chords of the circular curves shown in Fig. 9-12 and Fig. 9-13, compute the middle ordinates by both the exact and approximate methods.

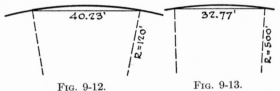

FIG. 9-12.   FIG. 9-13.

9-7. Tangent Curves. In addition to being tangent to straight lines, circular arcs are sometimes joined, the curves having a common tangent at the point at which they meet. Figure 9-14

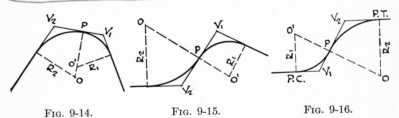

FIG. 9-14.   FIG. 9-15.   FIG. 9-16.

shows the arcs of two circular curves both of which are tangent to the line V_1-V_2, the centers of the arcs being on the same side of the tangent line; the resulting curve is called a *compound curve*. Figure 9-15 shows the arcs of two circular curves having the common tangent V_1-V_2, the centers of the arcs being on opposite sides of the tangent; this type of curve is a *reverse curve*. In designing roads and highways, reverse curves should be avoided whenever possible.

The method of establishing points on tangent curves is similar to the methods employed for simple curves. For conditions in which two curves meet at a common tangent, *a line joining the centers of the curves and the common point of tangency* (line $OO'P$ in Fig. 9-14 and line OPO' in Fig. 9-15) *is always a straight line and is always at right angles to the common tangent.*

In planning tangent curves the simplest procedure is to begin by drawing the tangent line, V_1-V_2, computing its length, and selecting point P, the common point of tangency. After this, the tangent distances are made equal and the radii of the curves computed as previously described.

The method of connecting two straight lines with a reverse curve is shown in Fig. 9-16. A line V_1-V_2 intersecting the two straight lines is drawn and its length is computed; this line will be the tangent common to the two curves. A point, such as P, is selected on line V_1-V_2, and through this point a line is drawn at right angles to V_1-V_2. The centers of the arcs of the curves will lie on this line. The tangent distances are now laid off on the two given lines; V_1-$P.C.$ = V_1-P and V_2-$P.T.$ = V_2-P. At points $P.C.$ and $P.T.$ lines are drawn at right angles to the two given lines. The intersection of these lines with the line through P perpendicular to V_1-V_2 determines points O' and O, the centers of the arcs; and thus the radii R_1 and R_2 are established.

CHAPTER 10

LEVELLING

10-1. Levelling. The *elevation* of a point is its vertical distance above or below a datum plane. *Levelling* is the process by means of which differences of elevation of two or more points are determined. The previous matter in this book has dealt with lines, angles, and areas assumed to be in the same plane. The surveyor, however, is also called upon to determine the rise and fall of the ground so that provision may be made for convenient access to buildings, for grading, and for proper drainage of the site.

The instruments used in levelling are the *level* and *levelling rod*. Because of its accuracy in manufacture as well as the length (about 18″) of its telescope, the *engineers' level* affords the greatest accuracy in results. The engineers' transit and builders' level are also used for taking levels, but they do not afford the accuracy obtained by the engineers' level.

Levelling rods are about 7′ in length and are extendable to twice their collapsed length. The face of the rod is usually marked in feet and tenths of a foot, and the vernier on the movable target permits a reading of one hundredth of a foot. Some rods have graduations that give readings of one thousandth of a foot.

10-2. Taking Levels. In order to determine the elevation of a point it is necessary to begin with some point the elevation of which is either known or assumed.

EXAMPLE

EXAMPLE. Figure 10-1 shows two points, *A* and *B*, the known elevation of point *A* being 202.58. Determine the elevation of point *B*.

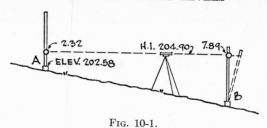

Fig. 10-1.

SOLUTION. STEP 1. At any convenient point the instrument is set up and levelled. This point need not be on a line between points A and B.

STEP 2. A reading is now taken on a rod held on point A. This reading is called a *backsight* or *plus* $(+)$ *sight*. The reading is 2.32, hence the *height of the instrument*, H.I., is $202.58 + 2.32$, or 204.90.

STEP 3. Next, the instrument is revolved and is sighted on a rod held on point B; the reading is 7.89. This reading is the *foresight*, or *minus* $(-)$ *sight*. Since the height of the instrument is 204.90, $204.90 - 7.89 = 197.01$, the elevation of point B.

The terms *backsight* and *foresight* are misleading since they do not indicate the direction in which the sights are taken. The backsight is the rod reading taken on a point the elevation of which is known. The foresight is the rod reading on the point the elevation of which is to be determined. The backsights are always plus, and the foresights are always minus.

In taking the levels of various points, the differences in level may be so great that the instrument must be moved from one position to another. When this is done, the instrument is sighted on a point the elevation of which has been found previously. Such a point is called the *turning point;* before its elevation was found, its rod reading was a foresight. As an example, the elevation of point B, Fig. 10-1, has been found to be 197.01. For a new position of the instrument, point B is the turning point and becomes the backsight for determining the elevation of other points.

10-3. **Accuracy in Taking Levels.** To determine elevations with accuracy, the instrument must be adjusted so that the bubble in the vial tube is in the exact center. In revolving the instrument the position of the bubble should be checked for each reading. Regardless of the direction in which the instrument is sighted, the line of sight must be in a horizontal plane.

Another item of importance is to have the levelling rod held *in a vertical position when the reading is taken.* If the rod is not vertical, as indicated by the dotted line in Fig. 10-1, a true reading cannot be taken. If the rod is not vertical the reading is always greater than the true reading.

10-4. Datum and Bench Mark. The theoretical level plane to which elevations are referred is called *datum;* it is usually mean sea level. A *bench mark* is a permanent point of which the elevation, above or below datum, is known. The U. S. Coast and Geodetic Survey and local authorities have established the elevations of numerous bench marks throughout the country, and from these points other bench marks may be established. The location and elevation of these official bench marks may be obtained from the engineering departments of the civil authorities. When showing elevations on a drawing a note should be given to indicate the datum that is used, U. S. Geodetic or local. When no authoritative bench mark is available in the vicinity of the proposed work, any convenient permanent point may be taken as a datum and assigned an elevation, usually 100. This point is the height on which all other elevations shown on the drawing are based. This procedure, however, should be followed only when an established bench mark is not available.

During the construction of a building it is generally desirable to establish a temporary bench mark on the site. This bench mark is used to determine the various elevations shown on the drawing. In locating such a bench mark care should be exercised to see that it is in a protected position and that it is not likely to be disturbed during the construction operations. Points on nearby buildings, curbstones, or other permanent objects are preferable to wooden stakes used as bench marks.

10-5. Errors Due to Curvature and Refraction. An error in levels is always present when the distance between the points is great. This error results from the curvature of the earth's surface and the refraction or bending of the line of sight as it passes through the atmosphere. In sights of ordinary length, the combined error is about 0.001′ in a distance of 200′, so small that it is neglected in computations. In taking levels *the error may be eliminated entirely by placing the instrument at equal distances from the back-sight and foresight.* This is called *balancing the sights;* the equal distances may be approximated by the eye.

10-6. **Correcting Errors in Levels.** In taking levels we begin with a bench mark of known or assumed elevation, preferably an *official bench mark.* Curb ele-vations, as shown on city or highway plans, should be used with caution since they may have settled after their installation. In order to check the accuracy of work, closed circuits returning to the orig-inal bench mark should be made whenever possible. In road surveys, sewer lines, etc., a closed circuit is impractica-ble. In such cases the levels should be tied in with other bench marks along the way, using them as turning points.

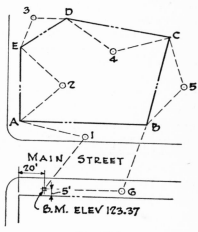

Fig. 10-2.

Figure 10-2 shows the various positions of the instrument in obtaining the elevations for the corners of the plot shown in Fig. 8-1. By consulting maps in the township engineer's office, the position of a bench mark was found to be on the south side of Main Street, as indicated on Fig. 10-2. It was described as a stone marker, and the elevation given was 123.37. This bench mark was located at the site and the instrument set up at point 1, a point so located that the backsight on *BM* and the foresight on point *A* were approximately equal distances. Note that the instrument need not be set on a line between the backsight and foresight. Readings on the two points were then taken, and the instrument was moved to point 2. From this position readings were taken on point *A* (the turning point) as a backsight and point *E*, the foresight. The procedure was continued around the plot, using the corners of the plot as turning points, and returning to the original bench mark. The result of the readings were re-

corded as shown in Fig. 10-3; the computed elevations are given
in the fifth column.

Note that, in returning to the bench mark, the computed elevation was found to be in error by 0.02′, (123.39 − 123.37). This
error should be distributed among all the points in proportion to
the distance travelled around the circuit. However, if this were
done, the elevations would be given in thousandths of a foot.
Such elevations would imply greater accuracy than could be obtained by the instrument. In this problem it is sufficiently accu-

FIELD NOTES

STA.	B. S. +	H. I.	F. S. −	ELEV.	CORRECTED ELEV.	REMARKS
B. M.				123.37	123.37	STONE MARKED IN S. SIDE OF MAIN ST.
A	5.18	128.55	3.98	124.57	124.57	
E	7.35	131.92	5.23	126.69	126.68	
D	10.78	137.47	6.71	130.76	130.75	
C	3.62	134.38	11.60	122.78	122.76	
B	4.80	127.58	6.32	121.26	121.24	
B. M.	6.85	128.11	4.72	123.39	123.37	

Fig. 10-3.

rate to drop the elevation of point E 0.01′ and to drop point C an
additional 0.01′, thus balancing the circuit. The corrected elevations are shown in the sixth column of the table, Fig. 10-3.

When levelling is done with an engineers' level the error in feet
should not exceed $0.05\sqrt{M}$, in which M is the length of the
circuit in miles. When a less accurate instrument is used, such
as an engineers' transit or builders' transit or level, the error to
be expected will probably be two or three times this amount.

When the sight distances are less than 200′, the errors resulting
from curvature and refraction are small and a saving of time may
be effected by taking fewer turning points and taking several
elevations from a single location of the instrument.

For example, suppose that in the illustration on page 119 we
had decided to set up the instrument at point 2 (Fig. 10-2), from
this point sight points A, E, and D, and then, using point D as a
turning point, set up the instrument at point 4 and sight points
C, B, and the bench mark. The field notes and computations

would have resulted in the tabulation shown in Fig. 10-4. In this instance the error of vertical closure is 0.03. To correct this discrepancy the height of the instrument at point 2 is reduced

FIELD NOTES

STA.	B. S. +	H. I.	F. S. −	ELEV.	CORRECTED ELEV.
B.M.				123.37	123.37
2	8.30	131.67			131.66
A			7.10	124.57	124.56
E			4.98	126.69	126.68
D (T.P.)			0.91	130.76	130.75
4	3.62	134.38			134.36
C			11.60	122.78	122.76
B			13.02	121.36	121.34
B.M.			10.98	123.40	123.37

FIG. 10-4.

0.01, at point 4 it is reduced 0.02, and the bench mark is reduced 0.03 to check with the original bench mark of 123.37.

10-7. Inverse Levelling. When it is necessary to determine the elevation of the underside of a bridge, a ceiling, or similar object, the following method may be used.

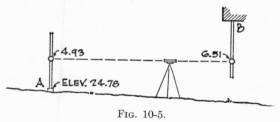

FIG. 10-5.

Suppose that we are given the elevation of point A, in Fig. 10-5, as 24.78 and are required to find the elevation of point B. The instrument is set up, and a reading is taken on the rod held on point A. This reading, 4.93, is a backsight or plus sight. The rod is now held on point B in an inverted position; that is, the zero mark on the rod is at the top. The reading is taken and is found to be 6.51. From the diagram it is seen that the elevation of point B is 24.78 + 4.93 + 6.51, or 36.22.

10-8. **Profiles.** Surveys of roads, railways, sewer lines, and other layouts of a linear character are often plotted in profile. _A profile is the vertical projection of the line of intersection of a vertical plane with the surface of the ground._ Profiles are usually taken at the center lines of roads and highways.

In plotting profiles it is customary to use an exaggerated vertical scale to show more clearly the changes in elevations. The

STA.	B.S. +	H.I.	F.S. −	ELEV.
B.M.	3.42			197.78
A	4.35		2.89	
B	5.29		3.71	
C	1.71		5.47	
D	2.64		4.21	
B.M.			1.13	

FIG. 10-6.

STA.	B.S. +	H.I.	F.S. −	ELEV.
B.M.	9.95			326.98
1				
A			3.26	
B (T.P.)	4.37		5.32	
2				
C			6.72	
D			4.58	
E (T.P.)	3.42		1.17	
3				
F			2.31	
B.M.			11.25	

FIG. 10-7.

STA.	B.S. +	H.I.	F.S. −	ELEV.
BM₁	2.21			78.03
A	5.73		3.24	
B	3.86		8.43	
C	1.73		5.21	
BM₂	8.26		3.61	71.07
D	4.92		6.98	
E	7.08		1.56	
BM₃	6.50		9.15	73.64
F	2.61		7.76	
G	5.56		4.36	
H	1.20		3.87	
BM₄			5.01	68.51

FIG. 10-8.

horizontal scale is generally five or ten times the vertical scale. For example, if the horizontal scale is $1'' = 40'\,0''$, the vertical scale might be $\frac{1}{8}''$ or even $\frac{1}{4}'' = 1'\,0''$. Specially prepared profile paper with ruled lines is of great convenience in plotting profiles. An example of a roadway profile is shown in Fig. 14-6.

Figure 10-6 shows the field notes, backsights, foresights, and bench mark for a four-sided plot. The corners of the plot, A, B, C, and D, are used as turning points. See Problem 10-8-A.

The field notes for a six-sided plot are shown in Fig. 10-7. Points A, B, C, D, E, and F are the corners of the plot. Points 1, 2, and 3 are positions of the instrument. Points B and E are used as turning points. See Problem 10-8-B.

Figure 10-8 shows the field notes for the linear survey of a sewer line. Points A, B, C, D, E, F, G, and H are points along the route whose elevations are to be determined. These points, as well as the bench marks, BM_1, BM_2, and BM_3, are used as turning points. See Problem 10-8-C.

PROBLEMS

10-8-A-B-C. Figures 10-6, 10-7, and 10-8, referred to above, show field notes for certain problems in levelling. For each set of field notes the figures balance out; there is no error of vertical closure. Compute the various instrument heights and the elevations of each point.

CHAPTER 11

CONTOURS

11-1. Contours. The usual map shows only two dimensions, length and breadth. Various devices are employed to indicate the third dimension or relative differences in elevation, but the most practical method is the use of *contours*. Often, the differences in elevation of a site may be more thoroughly understood by inspecting a contour map than by inspecting the site itself.

A contour is a line drawn on a map or plan which connects all points that have the same height above some reference plane. The

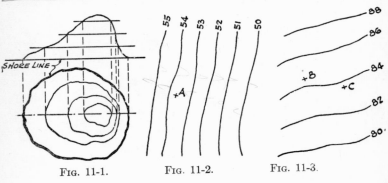

FIG. 11-1.   FIG. 11-2.   FIG. 11-3.

reference plane is the *datum plane*, and on many maps it is the mean sea level. *The vertical distance above the datum plane is the elevation.* A contour may be visualized as the intersection (shown in plan) of a level plane, such as the surface of a body of water, with the undulating surface of the ground. Shore lines of bodies of still water illustrate contour lines; a rise or fall of water creates other contour lines. This is illustrated in Fig. 11-1, which shows a small island in both plan and elevation. As the water rises new shore lines are formed and these lines, in plan, constitute contours.

11-2. Contour Intervals. A *contour interval* is the *vertical* distance between contours. A smaller contour interval will result in a greater number of contours on a map. The selection of the contour interval depends upon several factors: the purpose for which the map is to be used, the scale of the drawing, the roughness of the terrain, and the cost of obtaining the data required to plot the contours. On small-scale maps, contour intervals of 50′ and 100′ are often used. For site plans, however, where more detailed information is demanded, intervals of 5′, 2′ and 1′ are commonly employed. For building sites 1′ intervals are recommended.

When the contour interval has been decided upon, the same interval should be maintained for the entire drawing. More than one contour interval on a drawing frequently leads to errors in interpretation. When a certain detail requires more information than is afforded by the contours shown, *intermediate contours* are sometimes drawn between the regular contours; they should be drawn with a light dotted or dashed line and should extend only as far as the detail requires.

11-3. Determining Elevations from Contours. On an accurately drawn contour plan, the elevation of any point may be determined by interpolation. The contour interval for the contours shown in Fig. 11-2 is 1′, the contours 50, 51, 52, 53, 54, and 55 being shown. Suppose that we are asked to determine the elevation of the point A. This point lies about $7/10$ the distance from contour 53 to contour 54. Since each contour indicates a *vertical* distance of 1′, point A has an elevation of 53.7.

Figure 11-3 shows contours drawn with 2′ intervals, as indicated by their numbers. To find the elevation of point B we first note that it is located $3/10$ the distance from contour 84 to contour 86. As the contour interval is 2′, the elevation of point B is $84 + (3/10 \times 2)$, or 84.6. Point C lies at $8/10$ the distance from contour 82 to contour 84. The contour interval being 2′, the elevation of point C is $82 + (8/10 \times 2)$, or 83.6.

11-4. Significance of Contours. The contours of a map reveal definite characteristics of the terrain. A knowledge of these characteristics and their significance is essential in their interpretation.

1. Closely spaced contours at the higher elevations with greater spacings at lower levels indicate a concave slope. When the spacing is greater at the top of a slope and closer together at the

bottom, the slope of the ground is convex. These conditions are indicated in Figs. 11-4 and 11-5, respectively.

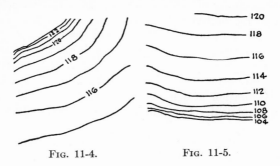

FIG. 11-4.   FIG. 11-5.

2. Evenly spaced contours indicate a uniform slope. On a plane surface the contours are straight, evenly spaced, and parallel.

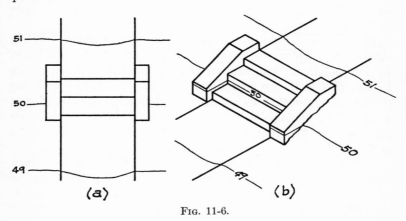

FIG. 11-6.

3. Every contour is a continuous line which *closes upon itself* somewhere on the earth's surface though not necessarily within the limits of a drawing. A contour cannot stop within the confines of a drawing; it must be a closed curve or, if it enters at the border of a drawing, it must leave the sheet at some other point on the border. Figure 11-6 (*a*) shows a plan of a pathway and steps with contour line 50 apparently stopping at the cheek wall of the steps. Figure 11-6 (*b*), a perspective, shows that this con-

tour line actually follows around the cheeks and a riser of the steps.

4. A closed contour surrounded by other contours indicates either a summit or a depression, the distinction being indicated by the contour numbering. Since no numbering is shown on contours A and B, Fig. 11-7, the contours might indicate either a

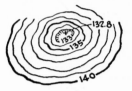

Fig. 11-7.   Fig. 11-8.

summit or a depression at the top of a slope. If contours A and B were numbered 136 and 137, respectively, we would know that a summit was indicated. The numbering of the contours in Fig. 11-8 indicates a depression at the bottom of a slope. To aid further in the identification of a depression, short lines drawn at right angles to a contour are sometimes used. These indications, called *hachures*, are shown at contour 133 in Fig. 11-8. The highest or lowest elevation is shown by a *spot elevation*, such as elevation 132.8 in Fig. 11-8.

5. Except for one condition, contour lines never cross each other, for such a condition would indicate a point having two different elevations. The exception is a vertical or overhanging cliff, as shown in both plan and profile in Fig. 11-9. For such a condition one contour must cross another at two points. Figure 11-9 shows the method of plotting a profile, the profile being taken on line A-A. Contours may appear to coincide at vertical excavations or at buildings. The perspective drawings, Figs. 11-10 (a) and 11-10 (b), show that the contours actually run around the face of the excavation or the face of a building.

6. Contour lines are perpendicular to lines of the steepest slope. Figure 11-11 shows two contour lines, 41 and 42. Since there is the same difference in elevation between point A and any point on contour 41, the steepest slope is found on the shortest line between the two contours. This line is a line at right angles to the contours. For the same reason, when contours cross ridge or valley lines they are perpendicular to them.

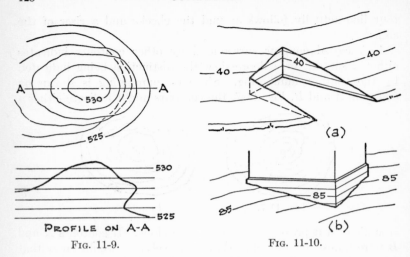

FIG. 11-9.

FIG. 11-10.

7. A stream and adjacent contour lines are shown in Fig. 11-12. When a contour line crosses a river or a stream, the contour first goes upstream, crosses the stream at right angles (the thread of the stream is the line of greatest slope) and then follows downstream.

8. The highest contours along ridges, and the lowest contours in valleys, always go in pairs. Figure 11-13 shows the contours

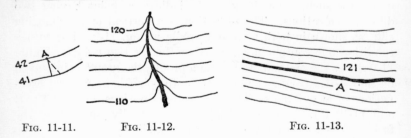

FIG. 11-11.   FIG. 11-12.   FIG. 11-13.

adjacent to a stream. If the lowest contour on one side of the stream is 121, then A, the lowest contour on the opposite side, must also be 121.

9. A contour never splits, as shown in Fig. 11-14 (a) and (b). A split could occur only when a knife-edged ridge or valley coin-

cided exactly with a contour line. Such a condition, of course, does not occur in nature. When sharp ridges or depressions do occur they would probably be represented as indicated in Fig. 11-14 (c) and (d).

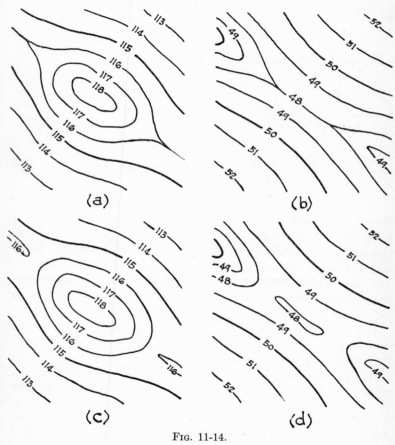

FIG. 11-14.

11-5. Plotting Contours. The best method of plotting contour lines for relatively small areas, such as building sites, is known as the cross-section or grid method. By the use of the transit and tape the plot of ground is first divided into a series of squares, called a grid; Fig. 11-15 shows such a grid. For the purpose of

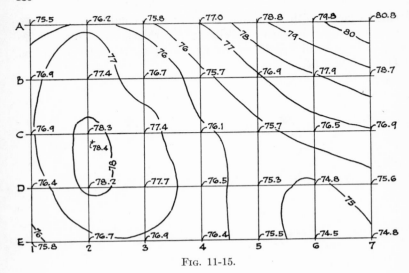

Fig. 11-15.

identifying specific points on the plot, the horizontal grid lines are lettered *A*, *B*, *C*, *D*, and *E*, as shown. The vertical lines are numbered 1 to 7, inclusive. As an example, with this system of notation the center of this particular plot is identified as *C*-4. The corners of the grid squares are marked with temporary stakes; the elevations of the ground at these points are taken and are so marked on the plan, as shown on Fig. 11-15.

The unevenness of the ground and the purpose for which the contour map is to be used determine the size of the grid squares; they vary from 10' to 100'.

Having laid out the grid and marked the elevations at the corners of the squares, our next task is to draw the contours. For the plot shown in Fig. 11-15 we will use 1' intervals. We note that the highest and lowest contour lines will be 80 and 75, respectively. The only contour line that will cross the corner of a grid square is 77; it intersects point *A*-4. Hence, we must now find the remaining points on the grid lines at which the contour lines will cross.

As an example, the elevations of points *E*-5 and *E*-6 are 75.5 and 74.5, respectively. Obviously, contour line 75 will intersect grid line *E* at the point halfway between grid lines 5 and 6. Next consider points D-5 and D-6, the elevations of which are 75.3 and 74.8. The difference in elevation between these two points is 75.3 − 74.8, or 0.5. Contour 75 lies between points D-5 and D-6, and, since 75 − 74.8 = 0.2, contour 75 will cross grid line D at $\frac{2}{5}$ the distance from D-6 to D-5. This, of course, is interpolation. It may be done graphically, the underlying principle being

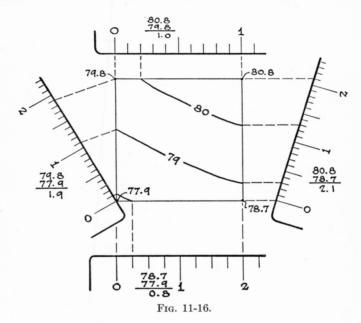

Fig. 11-16.

the division of a line into any number of equal parts by the use of a scale. Figure 11-16 is an enlarged drawing at the grid square shown in the upper right-hand corner of Fig. 11-15. Any convenient scales are used, their purpose being to find the points of intersection of contour lines 78, 79, and 80 on the grid lines. Scales are laid adjacent to the grid square, and the figure indicates the procedure of finding the contour points. Minute accuracy, however, is unnecessary, and an experienced topographer performs the interpolations mentally. When the contour points on all the grid lines have been determined the contour lines are drawn in *smooth* curves, as shown in Fig. 11-15.

PROBLEMS

11-5-A-B. The accompanying tabulations give the elevations of the grid intersections for two separate plot plans somewhat similar to the plan shown in Fig. 11-15. Lay out the grids with squares 1″ on a side and draw the contour lines with 1′ intervals.

	1	2	3	4	5	6
A	35.3	36.4	37.0	38.1	38.9	39.6
B	36.9	37.9	38.8	39.9	40.4	39.7
C	38.0	39.0	40.4	41.5	40.4	39.0
D	38.8	40.3	41.7	40.4	39.2	38.3
E	39.3	40.5	39.9	39.2	38.3	37.3
F	39.5	39.4	38.9	38.0	37.1	36.3

Elevations of Grid Intersections for Problem 11-5-A

	1	2	3	4	5	6	7
A	95.0	94.8	95.7	95.6	95.5	94.8	94.1
B	95.7	94.7	96.4	97.2	97.3	96.3	95.0
C	96.7	95.2	95.5	97.4	98.6	97.4	95.7
D	97.5	96.4	94.8	95.8	96.8	96.7	95.8
E	97.1	98.1	96.4	94.9	95.3	95.4	94.6
F	96.4	98.0	97.8	96.6	94.9	93.0	93.2
G	95.4	96.6	97.4	96.8	95.6	94.4	92.6

Elevations of Grid Intersections for Problem 11-5-B

CHAPTER 12

USES OF CONTOURS

12-1. Uses of Contours. After the contour map of the site has been drawn it can be used advantageously in determining the most desirable position for the buildings, the roadways, the paths, and the required grading. Invariably, grading must be performed, and this will result in the revision of certain contour lines. A study of the natural contours and the contours showing new grades enables us to compute the volumes of cut and fill that are required. See Chapter 13.

12-2. Grades. With respect to site engineering, the term *grade* has two different meanings. Very often the word *grade* is used synonymously with the word *elevation*, the height of the earth's surface at some specific point.

Another meaning of the word grade is *gradient* or *slope*. If we are given two points on a plan, one higher than the other, the grade is found by dividing the difference in their elevations by the horizontal difference (length) between them; that is,

$$G = D \div L \qquad \text{or} \qquad D = G \times L \qquad \text{or} \qquad L = D \div G$$

in which G = the grade (gradient).

D = the difference in elevation between the two points.

L = the horizontal length between the two points.

The grade is expressed as a decimal or as a percentage. As an example, Fig. 12-1 represents the plan of a roadway of uniform slope, the dot-and-dash line being the center line. Points A and B have elevations of 158.62 and 155.72, respectively, and the horizontal distance between them is 116′ 0″. The difference in elevation is 158.62 − 155.72, or 2.90′. Then $G = D \div L$, or

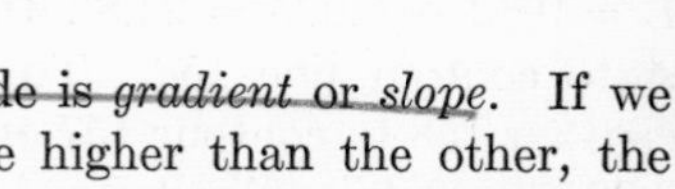

Fig. 12-1.

$G = 2.90 \div 116 = 0.025$, or 2.5%, the grade (slope) of the road-way. A common indication of a grade is shown on Fig. 12-1; *the arrow always points down the slope* and the grade is expressed as a decimal.

When the grade of a line is known, the elevation of any point on the line may be determined by use of the formula $G = D \div L$. Suppose, for example, we wish to know the elevation of a point on line AB, Fig. 12-1, that is $22'$ to the right of point A. Then $D = G \times L$, or $D = 0.025 \times 22$, and $D = 0.55$, the difference in elevation. Hence $158.62 - 0.55 = 158.07'$, the elevation of the point $22'$ to the right of point A. Similarly, we can find the elevation of a point $46'$ to the left of point B. $D = 0.025 \times 46 = 1.15$, and $155.72 + 1.15 = 156.87$, the elevation of the point.

A problem that frequently arises is to determine the point on a line that is intersected by a certain contour line. For example, find the point on line AB, Fig. 12-1, at which contour line 158 would cross. The difference in elevation between this point and point A is $158.62 - 158.00$, or $0.62'$. Then $L = D \div G$, or $L = 0.62 \div 0.025 = 24.8'$, the distance to the right from point A to contour line 158. To find the distance to the right from contour 158 to contour 157 on line AB, $D = 1'\ 0''$ and $L = 1.0 \div 0.025 = 40'\ 0''$. Similarly, contour 156 will be $0.28 \div 0.025$, or $11.2'$ to the left of point B.

There are three terms in the basic formula $G = D \div L$. When any two are known, the third term may be computed. The method of computing grades and locating points that have been explained in this article must be thoroughly understood, for similar problems occur many times in the succeeding examples relating to contours. In practice the computations are invariably made with the slide rule.

12-3. **To Lay Out a Roadway of a Given Grade.** Figure 12-2 shows the contour lines of a given site; the contour interval is $5'$. Suppose we are required to determine the center line of a road-way, connecting points A and B, whose maximum grade will not exceed 4%. Since the contour interval is $5'$, the difference in elevation on the contour lines is $5'$. Then $L = D \div G$, or $L = 5.00 \div 0.04$, or $125'$, the minimum length of roadway be-tween any two contour lines. Therefore, with point A as a center, lay off an arc having a radius of $125'$ to intersect contour 145;

the intersection is point C. In the same manner, with point C as the center, find point D on contour 150, and with D as a center locate point E on contour 155. Note that if a straight line 125' in length is drawn between points E and G (on contour 160), a part of the line would be almost level and the remaining portion would have a grade exceeding the 4% maximum. For such a condition, a half-contour (interval 2.5') is drawn between con-

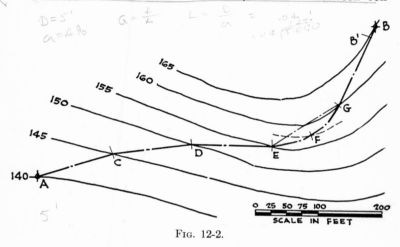

Fig. 12-2.

tours 155 and 160 and arcs EF and FG are laid off with radii of $125 \div 2$, or 62.5'. A curve is then constructed connecting points E, F, and G. Since the arcs are longer than the straight lines between points E, F, and G, the grades will be slightly less than the maximum 4%. An arc with a 125' radius intersects contour 165 at point B'. As line GB is slightly longer than GB', the grade of GB is somewhat less than the 4% maximum and thus the line indicating the center line of the roadway fulfills the requirements of the problem.

Various considerations may effect the final location of a roadway; the ideal solution is the roadway having the shortest length and requiring the least cut and fill. The roadway established in our solution started from point A. If we had selected point B as a starting point a slightly different route would have been obtained. For this type of problem widely divergent routes may

fulfill all the requirements but usually one solution is found to be
more economical than the others.

12-4. Establishing Grades for Surface Drainage. No natural
site is exactly level, and it is not desirable that any portion of a
finished (man-made) site be so. So-called level spaces, such as
athletic fields, lawns, or play areas, must have a slight slope to
permit rain water to run off. The earth surfaces adjacent to

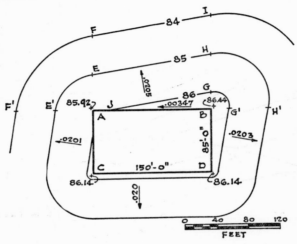

Fig. 12-3.

buildings should always be sloped so that surface water will drain
away from the building.

Figure 12-3 shows a rectangular building around which the soil
is to be graded to provide a minimum slope for the drainage of
surface water. The building shown is 150'x85', and spot elevations of finished surface elevations are given at the corners of the
building. These locations, indicated by crossed lines, are at the
corners but are shown a slight distance from the building for the
sake of legibility. The minimum grade (slope) of the earth surface will be assumed to be 2%. Our problem is to determine the
contours of the graded soil that will provide this slope *at right
angles to the faces of the building.*

We will begin at corner *A*, at which the finished elevation is
to be 85.92, by drawing a line from point *A* perpendicular to the

$D = .92$ $\quad a = \dfrac{b}{c}$
$G = .02$ $\quad L = 2\,9200$

AB face of the building; contour 85 will be somewhere on this line. The difference in elevation between the elevation at point A and contour 85 is $85.92 - 85$, or $0.92'$. Then $L = D \div G$ (Art. 12-2), and $L = 0.92 \div 0.02$, or $46'$. Hence, a point on the perpendicular line at $46'$ from point A is a point on contour 85; it is point E. In the same manner point E' is located $46'$ from the AC face of the building.

Next, let us find the point at which the line perpendicular to the AB face of the building intersects contour 84. The difference in elevation between contours 85 and 84 is $1'$. Then $L = D \div G$, and $L = 1 \div 0.02$, or $50'$, the distance from point E to contour line 84; this is point F. Similarly, $50'$ from E' locates point F'.

To find points G and G', $86.44 - 86 = 0.44'$. Then $L = D \div G$, or $L = 0.44 \div 0.02 = 22'$, the perpendicular distances from point B. Like the distance EF, distances GH, $G'H'$ and HI are all $50'$.

The same procedure is followed at corners C and D, and other points on the 86, 85, and 84 contours are thus found. Points having the same elevation are now joined to produce the contour lines. For example, points E and H are connected with a straight line because the earth surface between these points forms a level straight line. Contour line 86, GJ, is drawn parallel to EH through point G; it intersects the AB face of the building at point J. Distance AJ may be found as follows: The difference in elevation between points A and B is $86.44 - 85.92$, or $0.52'$, and the horizontal distance between them is $150'$. Then $G = D \div L$, or $G = 0.52 \div 150 = 0.00347$, the slope of the earth along line AB, as shown on Fig. 12-3. The difference in elevation between points A and J is $86 - 85.92$, or $0.08'$. Since $L = D \div G$, $L = 0.08 \div 0.00347$, or $23'$, the length of AJ.

At the corners of the building, circular arcs, $F'F$ for example, are drawn to complete the contours. Although the layout work is done mechanically, the contours of ground areas on final drawings are drawn free hand. When contour lines cross paths or paved roadways, they are drawn mechanically.

Note that the earth surface in contact with the CD face of the building is level, points C and D each being 86.14. For this reason the contours are parallel to the face of the building and are $50'$ apart; this results in the minimum required slope, 0.02.

On the other three faces of the building, however, the contours are not parallel to the face of the building and the slopes of the ground, *taken at right angles to the contours*, will be slightly in excess of the 2% minimum and will thus fulfill the requirements of the problem.

PROBLEMS

12-4-A. A rectangular building, 150'x83' in plan, is indicated in Fig. 12-4; spot elevations of finished grades are shown at the corners. Draw the building at a scale of 1" = 50' 0" and plot contours 85 and 86, the minimum slope away from and at right angles to the faces of the building being 2%.

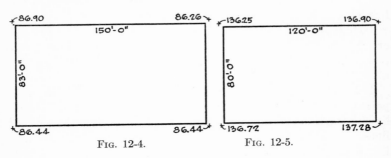

FIG. 12-4.   FIG. 12-5.

12-4-B. Figure 12-5 shows a building 120' long and 80' wide with spot elevations of finished grades at the corners as indicated. Draw the building at a scale of 1" = 40' 0" and plot contours 135, 136, and 137; the minimum slope of the ground away from and at right angles to the faces of the building will be 2%.

12-5. Sidewalk Grades Shown by Contours. It is essential that sidewalks and paths have a cross slope to permit rain water to drain off. The cross slope is the slope taken at right angles to the longitudinal boundary of the sidewalk. Local highway departments frequently specify that the cross slope must have minimum and maximum cross slopes of 2% and 3%, respectively. City plans generally show the established curb elevations, and these elevations are the starting points for computing the grades of the sidewalks.

Figures 12-6 (*a*) and (*b*) show two conditions of sidewalks adjacent to curbs. The curb elevations are given as data and are shown as spot elevations by small cross lines. For each condition shown, the length of the sidewalk is 115′ and the width is 15′.

Figure 12-6 (a) shows a sidewalk with a longitudinal 5% slope and a constant cross slope of 3% throughout its entire length. Beginning with the established curb elevations at points A and B, the elevations at the upper boundaries are computed; these are shown at points A' and B'. As an example, if the elevation of point A is 89.40, the width of the sidewalk is 15′, and the cross slope is 0.03, $D = G \text{x} L$, or $D = 0.03 \times 15 = 0.45′$.   Hence,

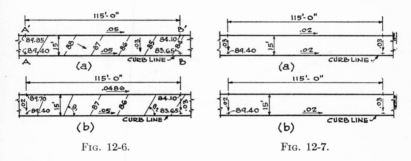

FIG. 12-6.   FIG. 12-7.

the elevation at point A' is $89.40 + 0.45$, or 89.85.   The points of *even contours* (84, 85, 86, etc., with no decimals) are located on the two boundaries of the sidewalk, and the contour lines are drawn connecting them. We know that two parallel lines determine a plane, and, since the longitudinal slopes at both boundaries are the same, the sidewalk is a plane surface and the contour lines are parallel. The rain water will drain from the sidewalk in a direction perpendicular to the contour lines. This is shown by the small arrow.

Figure 12-6 (b) shows a similar condition with the exception that the cross slope is 0.02 at one end and 0.03 at the other. This results in the two longitudinal boundaries not being parallel in space; the surface of the sidewalk is a warped surface, and the contour lines are not parallel. Note that angle θ_1 is greater than θ_2; hence *smaller cross slopes result in greater angles made by the contour lines with the curb line*. If there were no cross slope the contours would be perpendicular to the curb, an undesirable condition, for this would result in no cross drainage of surface water.

PROBLEMS

12-5-A-B. Figures 12-7 (*a*) and (*b*) represent sidewalks whose lengths between spot elevations are 115′ and whose widths to the curb line are 15′. For the slopes and curb elevations shown, compute the elevations at the upper and lower boundaries and draw the 88 and 89 contour lines.

12-6. Contours Drawn from Spot Elevations. For plane areas, it is possible to construct the contour lines when only spot elevations are known. When only spot elevations are shown on a drawing, it is assumed that the lines connecting them are of uni-

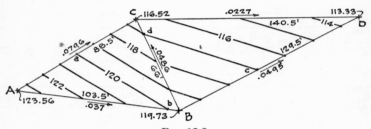

Fig. 12-8.

form slope. Spot elevations always take precedence over elevations determined by contours.

Figure 12-8 shows a four-sided plot with spot elevations given for points *A*, *B*, *C*, and *D*. Consider first the triangular area *ABC*. We know that *three points determine a plane* and that *contour lines are parallel on a plane surface*. Hence, if we establish one contour line on the area, the remaining contour lines may be drawn parallel to it. Suppose we are required to plot the contour lines on area *ABC*. Beginning with the side *AC* we compute the gradient (0.0796) and the position of the points of even contour. Next, the gradient on line *AB* is computed and *one* point of even contour is located, say point *b*, contour 120. Now, by drawing a line from point *b* to point *a* (contour point 120 on line *AC*) we establish contour line 120. Since we know that all contour lines on this plane surface *ABC* are parallel, we draw lines from the points of even contour on line *AC* parallel to contour 120, thus establishing all the contour lines on area *ABC*.

The four-sided plot shown in Fig. 12-8 is composed of two triangular areas, *ABC* and *BCD*; the contour lines on *ABC* have been established. To find the contours on the area BCD, begin by computing the gradient (0.0495) on line BD and locating the points of even contour. From contour point c on line BD draw a line to point d on line BC. This establishes contour 117. Through the points of even contour on line BD we now draw lines parallel to contour 117 and thus determine all the contour lines on area BCD.

A plot consisting of two or more triangular plane areas, as shown in Fig. 12-8, would exist only on paved surfaces. On a natural-earth or lawn surface erosion would not permit sharp angles, such as shown by the contours at line BC, to remain. The contours at such a line should be rounded.

12-7. **Contours on Fill.** In order to obtain the desired grading adjacent to terraces or other features of a building, it is often necessary to alter the original conformation of the terrain by cuts or fill. The extent of the work must be clearly indicated by the contour lines. Different procedures may be employed, but the following example illustrates an approved method.

EXAMPLES

EXAMPLE. The original ground area on which a terrace is to be constructed is a plane sloping surface, as indicated by the evenly spaced parallel contours in Fig. 12-9, the contour interval being 1' 0". The terrace, $ABCD$, is 25'x40' in plan; it begins on line AD, at which the original earth surface is 290.20, and has a downward pitch, toward edge BC, of 2% for drainage. The finished grading adjacent and at right angles to the sides of the terrace will have slopes of 1:5, that is, 1' fall for each 5' horizontally, a grade of 0.2. Show, by means of contour lines, the grading required to result in this condition.

SOLUTION. Since, by data, the elevations of the terrace at points A and D are each 290.20, the pitch is 2% and the width of the terrace is 25'; $D = G \times L$, or $D = 0.02 \times 25$, or 0.50'. Then the elevations at points B and C will each be 290.20 − 0.50, or 289.70. Thus it is seen that the only contour to cross the terrace area will be 290, and points a and a' will be $D \div G$, or $0.20 \div 0.02 = 10'$ from the AD side of the terrace. Contour 290 will be parallel to AD.

Next, we draw a line through point B perpendicular to BC. On this line we plot points on contours 289, 288, 287, and 286 of the finished grading. These points are b, c, d, and e, respectively. To find the distance Bb, $D = 289.70 - 289.00$, or $0.70'$. Then $L = D \div G$, or $L = 0.7 \div 0.2$ and $L = 3.5'$, the distance Bb. In a similar manner we find that $bc = cd = de = 5'$. Since BC is a level line, contours 289, 288, 287, and 286 will be parallel to BC.

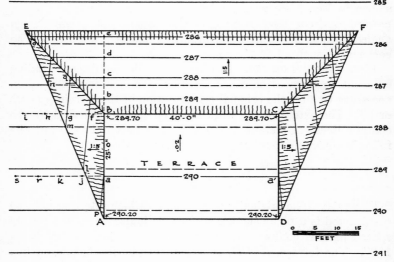

Fig. 12-9.

Therefore, we draw these contours (of indefinite length) through points b, c, d, and e.

With the same reasoning, we draw a line through B perpendicular to AB and locate points f, g, h, and i. $Bf = 3.5'$ and $fg = gh = hi = 5'$. But AB is not a level line, and therefore the new contours along side AB will not be parallel to AB. We know that two points determine a line. To find these required points draw a line perpendicular to AB from point a (a point on contour 290) and lay off $aj = jk = kr = rs = 5'$. Points j, k, r, and s would be points on the new contours, 289, 288, 287, and 286 *if* the grading were extended to these points. Therefore we connect points f and j, g and k, etc., and establish the positions of the new contour lines. These lines intersect the original contour lines at points *l, m, n,* and *o* and thus establish line *AE*. We know that *AE* and *BE* are straight lines because the intersection of two planes is a straight line. Observe that contour 289 (line *jf*) intersects the original contour 289 at point *l*, a point common to both the original and the new contour. Such a point is called a *point of no cut or fill;* it marks a point on the edge of surface *ABE*. This is true also of points *m, n, o,* and *p*. A line joining these points defines the intersection of the side slope with the original surface of the ground. Line *no* is extended until it meets line *BE* and establishes the lower edge of the forward slope. Note that line *BE* bisects angle *cqg*.

The contour lines on the right-hand side of the terrace are found in a similar manner. *Hachures*, short free-hand lines, are usually drawn to show the extent of the embankment. In the finished drawing the new contours and the undisturbed portions of the original contours are indicated by light solid lines. Those portions of the original contours effected by the grading are indicated by dash lines. In Fig. 12-9, as well as in the following figures, lines and points used to determine the positions of new contours are shown for the purpose of construction and explanation; they should not be shown on the final drawings.

EXAMPLE. The sides of terrace *ABCD*, shown in Fig. 12-10, are not parallel to the original contour lines. The earth fill at the sides of the terrace is to have a slope of 1:5, and the slope of the terrace is 2%. If point *D* (elevation 290.7) is kept at the elevation of the original ground surface, show the contours that will result in the required fill.

SOLUTION. We proceed by computing the elevations of points *B* and *C* (290.2) and determining the positions of the new contour lines by the method explained in the last example; the construction lines show the procedure. This, however, does not locate the position of point *E*. Line *AE* bisects the angle between the contours on the two adjacent slopes, but no contour lines occur in area *ADE*. But line *AD* is level, and any contours in the plane of *ADE* would be parallel to line *AD*. Consequently, if we bisect angle *daD* with the line *ap* we determine the *direction* of *AE*. Therefore, we draw a line through *A* parallel to *ap* and extend line

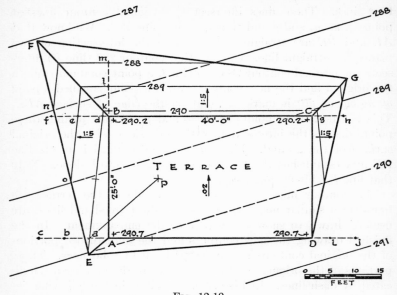

FIG. 12-10.

no; their intersection determines point *E*. In Fig. 12-10 to 12-17 inclusive the hachures have been omitted for the sake of legibility.

EXAMPLE. The terrace *ABCD*, shown in Fig. 12-11, is 20'x30', and line *AD* is level (elevation 290.20). The slope at the center line of the terrace is to be 3%, and, in addition, there is a cross slope of 1% both ways on line *BC*. The purpose of this is to reduce the tendency of surface drainage to erode the top of the fill along line *BC*. The earth fill at the sides of the terrace is to have a slope of 1:3. Determine the positions of the contours to provide the required fill.

SOLUTION. This problem differs from the former examples in that the contours of the original earth surface are not parallel lines, the ground is not a plane surface. In this example, the intersections of the fill with the original surface will not be straight lines. The first step is to compute the elevation of points *B* and *C*, 289.45. After this, the new contours are found by the methods previously described. The points at which the new contours

meet the original contours are connected by lines which mark
the boundary of the fill; in this instance they are irregular lines.

12-8. **Contours on Cuts.** On the inspection of a drawing in which
the contour lines indicate new grading, it may not at once be

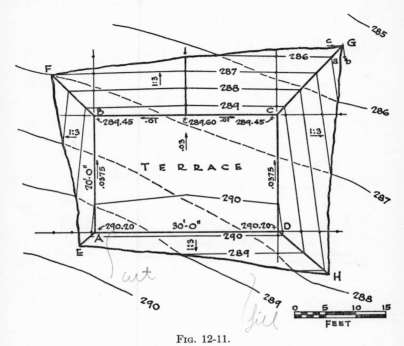

FIG. 12-11.

evident whether the work involved is a *cut* or *fill*. The following
method will be found to be helpful in their identification:

Turn the drawing so that you are apparently looking *downhill*.
This is indicated by the elevations of the original contour lines.

*If the new contours have been moved toward you, from their original
positions, the area is a cut. If the new contours have been moved
away from you, the area is a fill.* This system of identification may
be verified by referring to the figures discussed in Art. 12-7; they
are all examples of fill.

EXAMPLES

EXAMPLE.   The 20'x40' terrace *ABCD*, shown in Fig. 12-12,
is to have a 2% gradient toward edge *BC*.   Line *BC*, elevation
211.5, coincides with the natural grade.   The side slopes of the
cut will have a gradient of 1:2 *toward* the edges of the terrace.
Determine the contours to show the shape and extent of the cut.

SOLUTION.   The procedure explained in Art. 12-7 is followed,

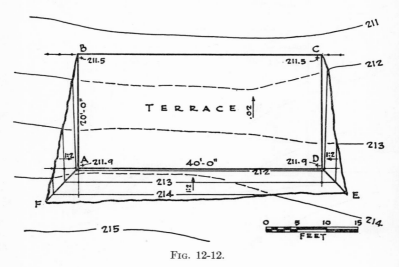

FIG. 12-12.

and the construction lines and contours are shown in Fig. 12-12.
This type of cut has the disadvantage of permitting surface water
to drain over the terrace.   To prevent this, the water may be
intercepted by a depression (gutter) at the bottom of the cut.
This is illustrated in the following example.

EXAMPLE.   Figure 12-13 shows a 20'x40' terrace, *ABCD*, with
a 3% gradient down toward edge *BC* (elevation 211.10), which is
level.   A V-type gutter is to be constructed at three sides of the
terrace.   The high point of the gutter, *F*, is 1.5' below *E*, a point
on the center line of the terrace.   The *bottom* of the gutter is to
have a 2% slope, and the side slopes of the gutter are to be 1:2.
Determine the positions of the contours to provide this condition.

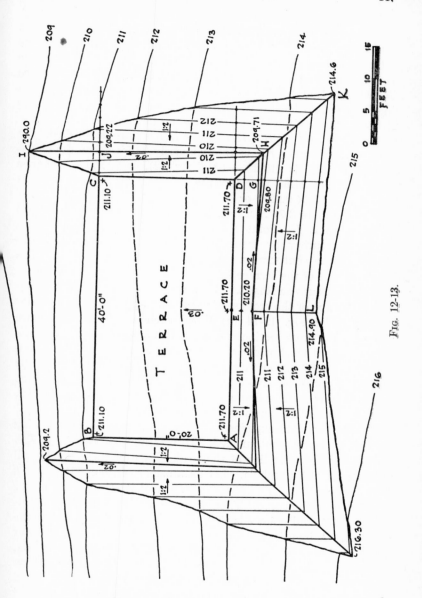

Fig. 12-13.

SOLUTION. The elevation of line AD is computed to be 211.70, and the elevation of point F will be $211.70 - 1.50$, or 210.20. Distance EF will be $1.5 \div 0.5$, or $3'$. Point G, the bottom of the gutter opposite D, scales approximately $20'$ from F; hence the elevation of G will be $210.20 - (20 \times 0.02)$, or 209.80. Distance DG will be $1.90 \div 0.5$, or $3.8'$. Point G is now plotted, and a line from F to G locates the bottom line of the gutter. Contours 210 and 211 may now be plotted on the terrace side of the slope to the gutter area $EFHICD$, and line DH is thus established. By scaling, the distance FH is approximately $24.5'$; hence the elevation of point H is $210.20 - (24.5 \times 0.02)$, or 209.71. Distance HJ also scales $24.5'$; therefore the elevation of point J becomes $209.71 - (24.5 \times 0.02)$, or 209.22. Distance CJ is $(211.10 - 209.22) \div 0.5$, or $3.76'$. Extended line HJ dies out on the natural grade at point I. Point I may generally be determined with sufficient accuracy by inspection. By the methods previously explained, the contours on area $FHIKL$ are now plotted. The left-hand side of the terrace is plotted in a similar manner.

12-9. **Contours on Roads.** When contours cross a road they repeat the shape of the cross section at what appears to be an exaggerated scale; Fig. 12-14 (a) shows such a condition. Figure 12-14 (b) is a cross section of the road showing a paved sidewalk on one side with a shoulder and V-gutter on the other. The crown of the road is $5''$ higher than the sides, and the cross section is the curve of a parabola. Note the gradients shown on the drawings. Point A, on the center line of the road, has an elevation of 592.62, and the gradient of the road is 0.028.

By the procedure explained in Art. 12-2, we begin by plotting the points of even contour, 592, 591, 590, and 589 (points a, b, c, and d) on the center line of the road. Each of these points will be the apex of a parabola. Let us consider contour 591; point b is on this contour. The crown of the road is $5''$, or $0.42'$ above the bottom of the curve. Then distance be will be $D \div G$, $0.42 \div 0.028$, or $15'$, and point e is thus plotted on the center line of the roadway. A line drawn through e at right angles to the center of the road establishes points f and g at the base of the parabola, and this permits us to draw curve fbg, which is contour 591 on the roadway. (The method of drawing a parabola is explained in Art. 14-1.) The height of the curb is $0.50'$; then, to find point h, distance $fh = D \div G = 0.50 \div 0.028 = 18'$. The sidewalk has a cross slope of 2%; hence the outer line of the sidewalk is $G \times L = 0.02 \times 10$, or $0.20'$ above the curb line. Then $hi = D \div G = 0.20 \div 0.028$, or $7.2'$. A line through i perpendicular to the curb line establishes point j, and line hj is contour

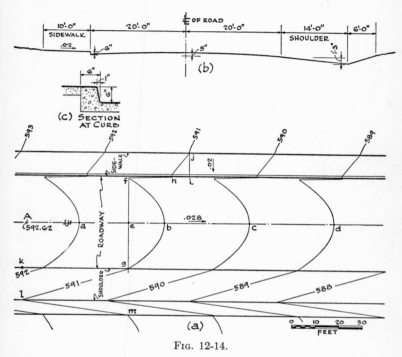

Fig. 12-14.

591 on the sidewalk. Point l, the bottom of the gutter, is located similarly. Distance $gk = D \div G = 1.25 \div 0.028 = 44.5'$, and point l is directly opposite k at the low point of the gutter. Point m is opposite g. This completes contour 591; the remaining contours are found in the same manner.

12-10. Cut and Fill at Sides of a Roadway. Figure 12-15 shows a straight roadway crossing contours that necessitate both a cut and fill. The road has a gradient of 0.04, and the side slopes of both the cut and fill, taken at right angles to the center line of the roadway, will be 1:4. Point A has an elevation of 352.80,

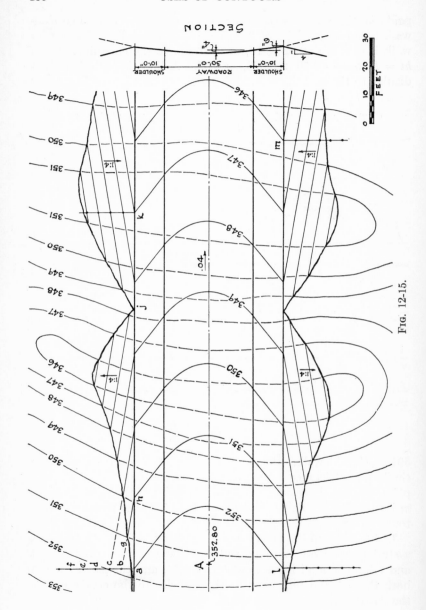

Fig. 12-15.

and the dimensions of the roadway and shoulders are indicated on the section.

The first step is to locate the points of even contours on the center line of the road and to plot the even contours on the road and shoulders, as explained in Art. 12-9.

At any point of even contour at the edge of the shoulder, for example, point a (elevation 352), draw a line perpendicular to the center line of the road. On this line lay off distances $ab = bc = cd$, etc. $= 4'$ (the slope is 1:4). At this slope point b will be 1' below point a and would be a point on contour 351 if the contour extended this far. Now point h at the edge of the shoulder is a point on contour 351; hence by joining points h and b we establish contour 351 on the slope. We note that this line intersects the natural contour 351 at point g. This is a point of no cut or fill, and it marks a point on the boundary of the fill. The remaining contours on the fill are drawn parallel to hb, and their intersections with the natural contours establish the extent of the fill. The fill dies out at point j the elevation of which, by inspection, is approximately 348.3.

To the right of point j the road goes into a cut. At point k, elevation 347, a vertical line is drawn and points $4'$ apart are laid off. Following the same procedure employed for the fill, the contours on the slope of the cut are drawn and the boundary line of the fill is established.

The contours showing the cut and fill on the opposite side of the road are drawn in the same manner, the vertical lines at points l and m indicate the method employed in finding their positions.

12-11. **Contours on Curved Roads.** In Art. 12-10 the portion of the road shown is straight, the adjacent banks are plane surfaces, and, consequently, the contour lines on the banks are parallel lines. When the road is curved the banks are curved surfaces and the contours are curved lines; the contours crossing the roadway are not true parabolas.

Figure 12-16 (a) shows a road that is curved between points A and B, the included angle (see Art. 9-1) is 60°, and the radius to the center line of the road is 100'. The width and gradients of the roadway and shoulders are shown in Fig. 12-16 (b). The gradient of the roadway *at the center line* is 0.025, and the banks of the fill have a slope of 1:4.

The first step is to compute the length of arc AB by the method explained in Art. 9-2. From Table 6 we find the length of the circular arc for 60° with a radius of 1 to be 1.04719; therefore,

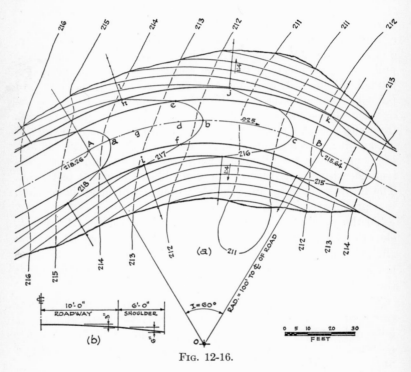

Fig. 12-16.

for a 100′ radius the arc length AB is 104.72′. Then the elevation of point B is 218.26 − (104.72 × 0.025), or 215.64. The points of even contours on the center line of the roadway, a, b, and c, are now found. For example, point a (on contour 218) is $D \div G$ = 0.26 ÷ 0.025, or 10.4′ from A. But this distance is on the *curved* line AB. When arc lengths are comparatively small they may be laid off as chords. In this case, however, we will use a more accurate method. The difference in elevation between points A and B is 2.62′; it is the difference that occurs in an angle of 60°. Then

$$\text{angle } Aoa = \frac{0.26}{2.62} \times 60 = 6°$$

$$\text{angle } aob = \text{angle } boc = \frac{1.00}{2.62} \times 60 = 23°$$

$$\text{angle } coB = \frac{0.36}{2.62} \times 60 = 8°$$

These computations may be made with the slide rule, reading the results only to the limit of accuracy that can be employed in using the protractor. Then, by use of the protractor, the above angles are laid off and points a, b, and c are established.

Now consider point b, the apex of the curve of contour 217. The crown of the road is 3″, or 0.25′ above the outer edge of the road. Then point d, the base of the curve, is $D \div G = 0.25 \div 0.025$, or 10′ from b. A line through points o and d intersects the edges of the roadway at points e and f. Distance $dg = D \div G = 0.5 \div 0.025 = 20′$, and a radial line through points o and g determines points h and i, points on contour 217 at the outer edge of the shoulders. Connecting points h, e, b, f, and i establishes contour 217 on the road and shoulders. The remaining contours on the road and shoulders are found in the same manner.

The next step is to draw radial lines through point o and points of even contours h, j, and k, and locate points 4′ apart to agree with a slope of 1:4. Points j and l are points on contour 216; hence they are joined by a curved line. The other contours are found in the same manner. The intersections of the contours with the original slope of the ground establish the boundary lines of the fill.

12-12. Banks at Sides of Steps. When steps occur in a path or walk, either a bank of earth or a retaining wall is required to accommodate the abrupt change in level of the adjoining earth surfaces.

Figure 12-17 (a) shows a 10′ wide path and steps, a section through which is shown in Fig. 12-17 (b). The banks at the sides of the steps are to have a slope of 1:3. On the right-hand side of the steps a line of indefinite length is drawn at right angles to the direction of the path. This line will be the top of the bank; it will have a 1% slope. Point A has an elevation of 49.12, and

point a (elevation 49) is plotted. It is seen that the top of the
bank will meet the natural slope of the ground at some point
whose elevation is between 48 and 49. Point c (elevation 48.5),
62′ from A, is plotted, but from the positions of the original contours it is apparent that the top line of the bank will not extend
this far. Point b (elevation 48.6), 52′ from A, is next plotted,
and we see that the top line of the slope meets the natural slope
at point B whose approximate elevation is 48.55. Distance $AB =$

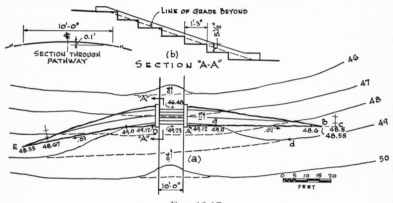

Fig. 12-17.

$0.57 \div 0.01$, or 57′. Points of even contours are located on the
cheek wall of the steps, and the contours on the bank are drawn
to intersect the natural contours at the bottom of the slope. Contour 49 runs out at the top of the bank at point a, which is above
the natural grade at this point. This requires that earth be filled
in here, and a portion of contour 49, ad, is drawn so that surface
water will be diverted from the path. Contour 50 is altered
similarly.

If the above method is followed on the left-hand bank, the
bank would be extremely long. To avoid this, the bank is bent;
it forms a warped surface, and the contours on it will be curved
lines. Suppose we limit the length of DE to 57′. The elevation
at D will be $49.12 - (57 \times 0.01)$, or 48.55. Then point E is
interpolated between contours 48 and 49, and line DE, 57′ in
length, is drawn in. Contours 47 and 48 on the bank are drawn
by the construction indicated on the drawing.

CHAPTER 13 — 5,6,7,8,9

COMPUTATIONS FOR CUT AND FILL

13-1. Cut and Fill. An ideal grading situation is that in which the volume of cut equals the volume of fill. On the completion of such an operation there remains no excess earth to be transported elsewhere and no additional material need be brought to the site. Such a condition can only be obtained by computing the volume of cut and fill before the work is begun.

13-2. Excavations for Buildings. When the excavation for a building is comparatively small and the surface of the ground has a uniform slope, the excavation may be considered to be a prism. By multiplying the area of the excavation by the *average* height of the corners we obtain the approximate volume of earth to be excavated.

EXAMPLES

EXAMPLE. A 12'x20' rectangular excavation, *ABCD*, is shown in Fig. 13-1. The contours show the natural surface of the ground to have an approximately uniform slope. Compute the approximate volume of earth to be excavated if the bottom of the excavation has an elevation of 92.5.

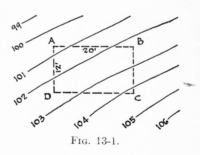

FIG. 13-1.

SOLUTION. By interpolating between the contours, the elevations at the corners are:

$$A = 100.7 \qquad B = 102.3 \qquad C = 104.2 \qquad D = 102.5$$

By subtracting the elevation of the bottom of the excavation, 92.5, from these elevations, the heights of the excavation at the

155

corners are:

$$A = 8.2' \qquad B = 9.8' \qquad C = 11.7' \qquad D = 10.0'$$

Then the average height multiplied by the area of the excavation is

$$\frac{8.2 + 9.8 + 11.7 + 10.0}{4} \times 12 \times 20 = 2{,}382\text{ft}^3 \quad \text{or} \quad \frac{2{,}382}{27}$$

$$= 88.2\text{yd}^3$$

When the excavations are more extensive, or when the surface of the ground is irregular, a somewhat similar system of computation is recommended. In this method the area is divided into a number of smaller areas, the average height is computed by taking the heights at a greater number of places, and the resulting volume is computed with greater accuracy.

EXAMPLE. The irregular area *ABCDEF*, shown in Fig. 13-2, represents the area of a proposed excavation of which the elevation of the bottom is 208.0. Compute the number of cubic yards of earth to be excavated.

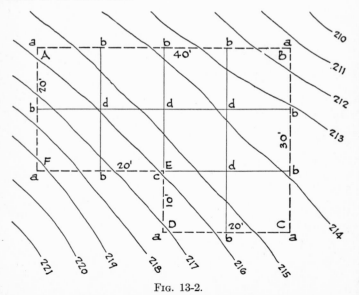

FIG. 13-2.

SOLUTION. STEP 1. Divide the area into a grid (a system of adjoining squares) consisting of squares 10′ on each side. Squares of any convenient size may be used.

STEP 2. Letter the corners of the squares so that the letter a designates the corners which occur in only one square, b designates the corners common to two squares, c designates the corners common to three squares, and d designates the corners common to four squares. See Fig. 13-2.

STEP 3. By the method used in the last example compute the height of the excavation at all the corners and also compute the sum of the heights of all the a's, b's, c's, and d's. The Greek letter Σ, used in mathematics, may be read "the sum of." Thus:

a's

215.7	211.3	214.6	217.2	219.3
208.0	208.0	208.0	208.0	208.0
7.7	3.3	6.6	9.2	11.3

Σa's = 38.1

b's

214.6	213.6	212.5	213.1	213.9	215.7	217.4	217.4
208.0	208.0	208.0	208.0	208.0	208.0	208.0	208.0
6.6	5.6	4.5	5.1	5.9	7.7	9.4	9.4

Σb's = 54.2

c's

215.9

208.0

7.9

Σc's = 7.9

d's

215.9	214.7	213.8	214.6
208.0	208.0	208.0	208.0
7.9	6.7	5.8	6.6

Σd's = 27.0

STEP 4. Compute the volume in cubic yards by use of the formula

$$\text{volume} = \frac{\text{area of 1 square}}{27} \times \frac{\Sigma a\text{'s} + 2\Sigma b\text{'s} + 3\Sigma c\text{'s} + 4\Sigma d\text{'s}}{4}$$

or,

$$\text{volume} = \frac{10 \times 10}{27} \times \frac{38.1 + (2 \times 54.2) + (3 \times 7.9) + (4 \times 27.0)}{4}$$

$$= 258$$

which is the number of cubic yards of earth to be excavated. On large or complicated areas this method will prove to be long and tedious; the procedure explained in Art. 13-5 and 13-6 will be found more efficient.

13-3. **Volume of a Prismoid.** A *prismoid* is a solid with parallel but unequal ends or bases whose other faces are quadrilaterals or triangles. In computing volumes of earth, various cross sections of cut or fill are considered; they indicate parallel areas at definite distances apart. The volumes bounded by these areas approximate prismoids and are so considered in the computations. Two methods are used for computing the volume of a prismoid.

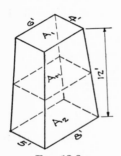

FIG. 13-3.

(*a*) AVERAGE END-AREA METHOD. Figure 13-3 shows a prismoid of given dimensions. The top and bottom faces, A_1 and A_2, are parallel planes. A formula commonly used for finding the volume of a prismoid is

$$\text{volume} = \frac{A_1 + A_2}{2} \times l$$

in which A_1 and A_2 are the areas of the two parallel faces and l is the normal distance between them, $(A_1 + A_2) \div 2$ being the *average end area*. The volumes obtained by use of this formula are not exact but generally are somewhat in excess of the exact values. Because of its simplicity, however, this formula is used extensively. Using it for the solid shown in Fig. 13-3,

$$\text{volume} = \frac{(4 \times 6) + (5 \times 8)}{2} \times 12 = 384 \text{ft}^3$$

(b) PRISMOIDAL FORMULA. The formula that gives the exact volume of a prismoid is

$$\text{volume} = \frac{A_1 + 4Am + A_2}{6} \times l$$

in which Am is the area of the section midway between the two parallel faces, the remaining terms being similar to those in the previous formula. Thus, for the prismoid shown in Fig. 13-3,

$$\text{volume} = \frac{(4 \times 6) + (4 \times 4.5 \times 7) + (5 \times 8)}{6} \times 12 = 380 \text{ft}^3$$

13-4. **Volumes of Cut or Fill Found from Contours.** A simpler and more efficient method of computing volumes of cut and fill

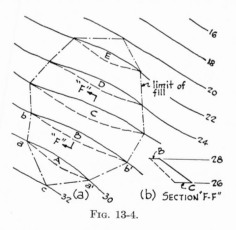

FIG. 13-4.

is to use the contours. In fact, this is one of the principal reasons for showing contours. Figure 13-4 (a) shows a portion of a plot, the contours indicating a fill. The dotted lines show the original contours and the solid lines the contours after the fill has been placed. Points a, a', b, b', etc., indicate the extremities of the fill. The areas between the original and finished contours are identified as A, B, C, D, and E. Figure 13 (b) shows a section through the fill at F–F, the dotted line being the original slope of the ground. Note that B and C are parallel planes, parallel faces of a solid the height of which is the vertical distance between them, actually the contour interval. If we consider the curved contours as a series of short straight lines, the earth included in the solid is, approximately, a prismoid. Therefore, if we compute the areas of B and C we can apply the "average end-area method" to compute the volume of the solid. Point c on contour 32 is a "point of no cut or fill," and a section, similar to F–F, taken through point c, is a triangle, the solid in this portion of the fill being approximately pyramidal in shape, the base being area A. This same condition is found between contours 20 and 22. The volume of a pyramid is $\frac{1}{3}$(base × height). Thus, the approximate volume of the entire fill may be found by adding together the volumes of all the individual prismoids and pyramids.

Let A, B, C, D, and E = the areas between the original and finished contours, in square feet; let i be the contour interval, in feet; and let V be the approximate total volume of the cut or fill, in cubic feet. Then

$$V = \frac{Ai}{3} + \frac{A+B}{2}i + \frac{B+C}{2}i + \frac{C+D}{2}i + \frac{D+E}{2}i + \frac{Ei}{3}$$

or

$$V = i\left(\frac{5A}{6} + B + C + D + \frac{5E}{6}\right)$$

From the preceding discussion it is seen that volumes computed by this method are only approximately correct, the contours themselves being only approximately exact. Thus, taking $\frac{5}{6}$ of the top and bottom areas (A and E) may be considered to be an unnecessary refinement. If these fractions are omitted, the rule that may be used to find the approximate volume is: *Add together all the areas, A, B, C, etc., and multiply their sum by the contour interval.* If the volume thus found is in units of cubic feet, divide by 27 to find the volume in cubic yards.

13-5. **The Planimeter.** There are several methods of computing the contents of an area bounded by an irregular line. The method of dividing the area into trapezoids, explained in Art. 7-2, can be used, or cross-section paper might be employed. But both of these methods are tedious, and the best procedure is to use an ingenious instrument called a *planimeter*, Fig. 13-5.

The planimeter is an instrument used for measuring areas of plane figures. It consists of a metal frame to which are attached

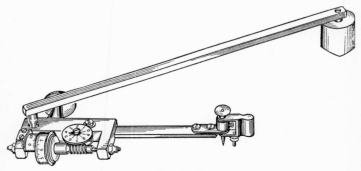

FIG. 13-5.

a weighted *anchor point*, a *measuring wheel* with counter and vernier, and a *tracing point*. To measure an area, the anchor point is pressed into the drawing board *outside* the area. The tracing point is placed at some marked point on the boundary of the area, and a reading of the counter and vernier is taken. Next, the tracing point is traced around the outline of the area *in a clockwise direction*, returning *exactly* to the starting point. Another reading is now taken, and the difference in the readings gives the area of the figure *in square inches*. Careful use of the instrument will result in an error not exceeding 1%.

The final step is to convert the square inches to the scale of the drawing. As an example, suppose the area found is 2.56in^2 and the scale of the drawing is $1'' = 40'0''$. Then $2.56 \times 40^2 = 4,096$ft^2, the area.

13-6. **Computing the Volumes of Cut and Fill with the Planimeter.** Figure 13-6 shows a portion of a site plan with the contours indicated in the conventional manner. The left-hand portion of the required grading is a fill and the portion on the right is to be a cut. The areas of cut and fill are found by use of the planimeter

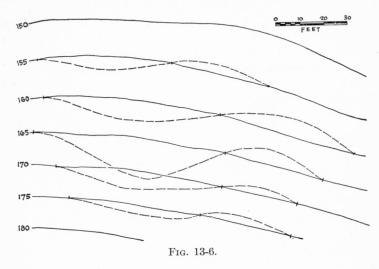

FIG. 13-6.

and the results are tabulated in Fig. 13-7. Note that this tabulation conforms with the formula given in Art. 13-4. The contour

CONTOUR	FILL	CUT
155	$\frac{5}{6} \times .13 = .11$	$\frac{5}{6} \times .08 = .07$
160	.27	.18
165	.47	.11
170	.21	.05
175	$\frac{5}{6} \times .09 = .08$	$\frac{5}{6} \times .06 = .05$

1.14 sq.in.	.46 sq.in.
(SCALE 1"=40'-0") ×1600 = 1824 sq.ft.	×1600 = 736 sq.ft.
(CONTOUR INTERVAL) ×5 = 9120 c.f.	×5 = 3680 c.f.
÷ 27 = 338 c.y.	÷ 27 = 136 c.y.

FIG. 13-7.

interval shown in Fig. 13-6 is 5'. Smaller contour intervals give greater accuracy in the results. When the dimensions of the site permit, a 1' contour interval is recommended.

13-7. Balancing Cut and Fill. The computations given in Fig. 13-7 show that there is a greater volume of fill than cut. If fill cannot be obtained from other parts of the plot, the contours should be revised so that there are approximately equal volumes of cut and fill. This may require several trial locations of contours or even necessitate raising or lowering a building.

13-8. Shrinkage and Settlement. When earth is excavated and placed in another position, it will occupy a space somewhat smaller than its original volume. The reason for this is the unavoidable loss of soil in transportation and, on steep slopes, the loss of material that is washed away by rain. Freshly deposited earth is compacted by rainfall, rollers, and tamping. Because of this, when balancing the volumes of cut and fill, an excess of fill, 5% to 10%, is commonly provided; that is, 100ft^3 of cut and 107ft^3 of fill would about balance.

13-9. Building Excavations Determined by the Planimeter. When an excavation has vertical sides (the usual condition), its volume may be included, in computing the volume of cut and fill on a plot, by using the planimeter. *Contours that meet the vertical sides of an excavation lie in the vertical face of the excavation on the uphill side.* This is shown in Fig. 11-10 (a).

EXAMPLE

EXAMPLE. Figure 13-8 (a) shows a 40′x60′ excavation, $ABCD$, the elevation of the bottom of the excavation being 80.5. As shown by the contours, the grading work to be performed outside the rectangle $ABCD$ is all fill and within the rectangle we have excavation, a cut. The scale of the plot drawing is $1'' = 32'\ 0''$. Compute the volumes of cut and fill shown by the contours and excavation area.

SOLUTION. Since the scale of the drawing is $1'' = 32'\ 0''$, 1in^2 of area registered by the planimeter is 32×32 or 1,024ft^2, the scale factor. The various areas and the required computations are tabulated in Fig. 13-8 (b), the procedure followed being that explained in Art. 13-4.

In computing the volumes of cut and fill on contour 89, areas $abca$ and $dDefd$ are tabulated under "fill" and area $cdAc$ (the contours lie in the vertical face of the excavation) under "cut."

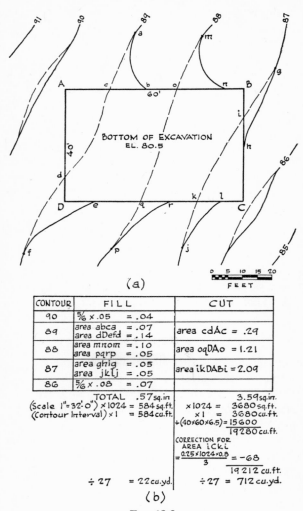

(a)

CONTOUR	FILL		CUT
90	⅚ × .05	= .04	
89	area abca	= .07	area cdAc = .29
	area dDefd	= .14	
88	area mnom	= .10	area oqDAo = 1.21
	area pqrp	= .05	
87	area ghig	= .05	area ikDABi = 2.09
	area jkīj	= .05	
86	⅚ × .08	= .07	

TOTAL .57 sq.in   3.59 sq.in.
(Scale 1"=32'-0") ×1024 = 584 sq.ft.   ×1024 = 3680 sq.ft.
(Contour Interval) ×1 = 584 cu.ft.   ×1 = 3680 cu.ft.
                        +(40×60×6.5)= 15600
                            19280 cu.ft.

CORRECTION FOR AREA iCki
$= \dfrac{0.25 \times 1024 \times 0.8}{3} = -68$
               19212 cu.ft.

÷27 = 22 cu.yd.   ÷27 = 712 cu.yd.

(b)

FIG. 13-8.

Cut and fill on contour 88 are recorded similarly.   On contour 87, areas *ghig* and *jklj* are fills and *ikDABi* is a cut.   The lowest contour intercepting the excavation is 87.   Since the bottom of the excavation has an elevation of 80.5, the bottom of the excavation is $87 - 80.5$, or 6.5′ below contour 87.   Therefore, the volume of excavation not previously accounted for is a prism of earth $40 \times 60 \times 6.5$, or 15,600ft^3, as shown in Fig. 13-8 (*b*).   This method of computation, however, includes an excessive volume; it is the volume *iCki* that is below the 87 contour level.   By interpolation, the original elevation of point C is 86.2.   Consequently, the excessive volume may be considered as a pyramid having *iCki* as a base $(0.25 \times 1,024)$ft^2 and a height of $87 - 86.2$, or 0.8′.   Thus, its volume is $\dfrac{0.25 \times 1,024 \times 0.8}{3}$, or 68ft^3.   Note that 1,024 is the scale factor for converting square inches on the drawing to square feet.   This volume, 68ft^3, is deducted from the volume of cut previously computed.   It is shown in the tabulation, Fig. 13-8 (*b*).

The volumes of cut and fill are 712yd^3 and 22yd^3, respectively. As noted in Art. 13-4, the volumes computed by this method are approximate.   If the excavation is a relatively large part of the total cut, the procedure explained in Art. 13-2 should be used for determining the volume of cut within the excavation lines.

CHAPTER 14

VERTICAL CURVES

14-1. **Parabolic Curves.** When the slope of a roadway changes and the difference between the grades exceeds 1% (½% on important and high-speed roads), the abrupt change is eased by the use of a vertical (profile) curve. Vertical curves are used when a downward slope changes to an upward slope (a sag curve) or

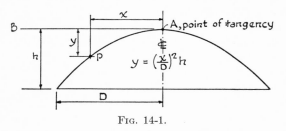

B

A, point of tangency

$$y = \left(\frac{x}{D}\right)^2 h$$

FIG. 14-1.

when an upward slope changes to a downward slope (a peak curve). Owing to the fact that they are readily computed and plotted, the vertical curves used for roadways are invariably parabolas.

The curve shown in Fig. 14-1 is a *parabola*. It is tangent to line AB at point A. The vertical height is h, and D is the half-width of the parabola. P is any point on the curve; its horizontal distance from the point of tangency is x, and y is its vertical distance from the tangent line. For any distance x, the distance y may be computed by the formula

$$y = \left(\frac{x}{D}\right)^2 h$$

which, in effect, states that *the vertical offsets are proportional to the squares of their distances from the point of tangency.* In using

166

this formula *be certain that all the terms are of the same units, feet or inches.*

EXAMPLES

EXAMPLE. In the parabola shown in Fig. 14-1 what is the offset y if $D = 20'$, $h = 8''$, and $x = 13'$?

SOLUTION. Since $h = 8''$, $h = (8/12)'$, or $0.67'$. Then, substituting in the formula,

$$y = \left(\frac{x}{D}\right)^2 h \qquad y = \left(\frac{13}{20}\right)^2 \times 0.67 \qquad \text{and} \qquad y = 0.28' \text{ or } 3\tfrac{3}{8}''$$

EXAMPLE. In the parabola shown in Fig. 14-2, h is the height of the parabola and D is the half-width. If D is divided into ten equal parts, compute the values of the y distances in terms of h.

SOLUTION. Beginning at the center line of the curve, the successive values of x are $D/10$, $2D/10$, $3D/10$, etc. Let the y distances corresponding to the various x distances be y_1, y_2, y_3, etc. Then, substituting in the formula for the parabola,

$$y_1 = \left(\frac{D/10}{D}\right)^2 \times h = \left(\frac{1}{10}\right)^2 \times h = \frac{1}{100} h$$

$$y_2 = \left(\frac{2D/10}{D}\right)^2 \times h = \left(\frac{2}{10}\right)^2 \times h = \frac{4}{100} h$$

$$y_3 = \left(\frac{3D/10}{D}\right)^2 \times h = \left(\frac{3}{10}\right)^2 \times h = \frac{9}{100} h$$

etc.

The remaining values of y are shown in Fig. 14-2. This figure shows the y distances when D is divided into ten equal parts. It will be found to be of great assistance in plotting any parabolic curve. Eleven points on the curve are shown, but usually a smaller number is sufficient to lay out the curve.

EXAMPLE. A roadway is $20'$ in width, the profile taken on a section *across* the road is a parabolic curve, and its height at the center line is $6''$. It is desired to lay out a template for this cross section. Compute the positions of various points on the curve.

SOLUTION. Since the road is 20′ in width, a horizontal line is drawn 10′ in length, corresponding to D in Fig. 14-2. On this line points 1′ 0″ apart are laid off. Then the y distances for these points are computed by using the values of y shown in Fig. 14-2.

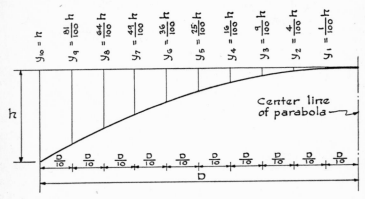

FIG. 14-2.

Note that $h = 6''$, or $0.5'$. Then $y_1 = \frac{1}{100} \times 0.5 = 0.005'$, $y_2 = \frac{4}{100} \times 0.5 = 0.02'$, and $y_3 = \frac{9}{100} \times 0.5 = 0.045'$. Similarly, $y_4 = 0.08'$, $y_5 = 0.125'$, $y_6 = 0.18'$, $y_7 = 0.245'$, $y_8 = 0.32'$, $y_9 = 0.405'$, and $y_{10} = 0.5'$.

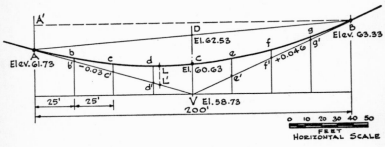

FIG. 14-3.

14-2. **Plotting Vertical Curves in a Roadway.** A downward slope of a roadway has a grade of 0.03; it intersects an upward slope having a grade of 0.046 at point V, as shown in Fig. 14-3. When traversing a road, a downward slope is marked minus (-0.03)

and an upward slope is marked plus (+0.046). The point of
intersection of the two grades, called P.I., is point V; its elevation
is 58.73. A vertical sag curve, 200' long in horizontal projection,
is to be placed in the road from points A to B. Suppose we are
required to determine the elevations of points on a parabolic
curve, joining points A and B, at 25' (horizontal) intervals.

ELEV. @ A = 58.73 + (100 × .03) = 58.73 + 3.00 = 61.73
@ b' = 58.73 + (75 × .03) = 58.73 + 2.25 = 60.98
@ c' = 58.73 + (50 × .03) = 58.73 + 1.50 = 60.23
@ d' = 58.73 + (25 × .03) = 58.73 + .75 = 59.48
@ V = 58.73 (given)
@ e' = 58.73 + (25 × .046) = 58.73 + 1.15 = 59.88
@ f' = 58.73 + (50 × .046) = 58.73 + 2.30 = 61.03
@ g' = 58.73 + (75 × .046) = 58.73 + 3.45 = 62.18
@ B = 58.73 + (100 × .046) = 58.73 + 4.60 = 63.33

$$@ \ D = \frac{61.73 + 63.33}{2} = 62.53 \quad \text{(Point D is midway between A and B)}$$

$$@ \ C = \frac{62.53 + 58.73}{2} = 60.63 \quad \text{(Point C is midway between D and V)}$$

VC = 60.63 − 58.73 = 1.90
$$b\,b' = g\,g' = (\tfrac{1}{4})^2 \times 1.90 = \tfrac{1}{16} \times 1.90 = .12$$
$$c\,c' = f\,f' = (\tfrac{1}{2})^2 \times 1.90 = \tfrac{1}{4} \times 1.90 = .48$$
$$d\,d' = e\,e' = (\tfrac{3}{4})^2 \times 1.90 = \tfrac{9}{16} \times 1.90 = 1.07$$

ELEV @ b = 60.98 + .12 = 61.10
@ c = 60.23 + .48 = 60.71
@ d = 59.48 + 1.07 = 60.55

@ e = 59.88 + 1.07 = 60.95
@ f = 61.03 + .48 = 61.51
@ g = 62.18 + .12 = 62.30

FIG. 14-4.

All measurements of length on a plan are horizontal projections. Thus, in Fig. 14-3, the lengths of ADB, ACB, and AVB
when shown in plan are all equal to the horizontal projection
$A'B$. In a vertical curve, $AV = VB$ in horizontal projection and
point C is always midway between points D and V. The elevation
of point V is given to be 58.73. The grades of AV and VB are
known, and thus we can compute the elevations of points A, B,
C, D, b', c', d', e', f', and g'; the computations are shown in Fig.
14-4. The elevation of point D is midway between the elevations
of points A and B, and point C is midway between D and V.
Lines AV and VB are tangent to the parabola at points A and B;

consequently the offsets, $b'b$, $c'c$, $d'd$, etc. (the y distances), are proportional to the squares of their distances from point A, the point of tangency. Point b' is $25/100$ or $\frac{1}{4}$ of the distance from point A to point V, and $VC = 1.9'$ (Fig. 14-4). Then $b'b = (\frac{1}{4})^2 \times 1.9 = 0.12'$. The elevation of b' is 60.98; hence 60.98 + 0.12 = 61.10, the elevation of point b. The elevations of other points on the curve are found in a similar manner and are shown in Fig. 14-4.

14-3. **Peak Curves.** When a roadway goes over the crest of a hill the vertical curve is inverted from the curve shown in Fig. 14-3. To find the elevations of various points on the curve, the

Fig. 14-5.

elevations of points on the grade lines AV and VB (Fig. 14-5) are found and the offsets are then *subtracted*, the procedure being similar to that described in Art. 14-2.

14-4. **High and Low Points on Vertical Curves.** When a drain or catch basin is to be placed at the *low point* of a curve on a roadway, it becomes necessary to locate the exact position of the low point. The horizontal distance from the point at which a curve begins to the low (or high) point of the curve is

$$\frac{lg_1}{g_1 - g_2}$$

in which l is the horizontal projection of the total length of the curve and g_1 and g_2 are the grade percentages at the beginning and end of the curve, respectively. In using this formula, refer to Art. 14-2 concerning plus and minus grades.

In Fig. 14-3 point C would be the low point of the curve only when AV and VB were of equal slopes. In this instance the low point is obviously to the left of point C, let it be point L. Substituting in the formula,

$$\text{horizontal projection of distance } AL = \frac{200 \times (-0.03)}{(-0.03) - (+0.046)}$$

$$= \frac{200 \times (-0.03)}{-0.076} = 79'$$

To find the elevation of point L, first find the elevation of point L'. Then, since the elevation of A is 61.73,

$$61.73 - (0.03 \times 79) = 59.36 \quad \text{the elevation of } L'$$

To find distance $L'L$ we use the principle given in Art. 14-1 and distance $L'L = (79/100)^2 \times 1.9 = 1.19'$. Therefore, $59.36 + 1.19 = 60.55$, the elevation of point L, the low point of the curve.

14-5. Profiles. With respect to surveying, a profile is the vertical projection of the intersection of a vertical plane with the surface of the earth. In plotting contours which cross highways, profiles are used advantageously. Since the slopes of roads are relatively small in comparison with their lengths, the vertical scale of a profile is generally exaggerated, being made five or ten times larger than the horizontal scale. Profiles are useful in checking vision or sight distances at the crest of peak curves. For highway work this is a necessity and is advisable for private roads and drives. Specially ruled paper for this type of work is available.

14-6. Plotting Contours by Use of a Profile. To explain the method of plotting contours crossing a roadway in which there is a vertical curve, consider the 200' curve shown in Fig. 14-3. A cross section through the roadway and shoulders is shown in Fig. 14-6 (a). Note that the road has a width of 40' and that 10' is the width of the shoulders. For drawing the profile, a larger scale has been used for the vertical dimensions; the heights are shown on Fig. 14-6 (c). The roadway with the shoulders is shown in plan in Fig. 14-6 (b).

The vertical curve shown in Fig. 14-3 is the curve at the center line of the roadway. This curve, ALB, is shown again *in profile* in Fig. 14-6 (c) by the solid curved line. The elevations of points A, B, and L are known, and the horizontal lines numbered 58 to 65 represent the *levels* of the various contour elevations. These contour levels are 1' apart and are drawn at a larger scale as an aid in plotting the contours. In Fig. 14-6 (a) we see that the sides of the road are 5", or 0.42', below the center line of the road and that outer edges of the shoulders are 13", or 1.08', below the center line. Now draw a vertical line through any point, such as *a*, on the curve shown in Fig. 14-6 (c), and lay off *ab* = 0.42' and *ac* = 1.08'. Thus, points *b* and *c* are points on the curves of the

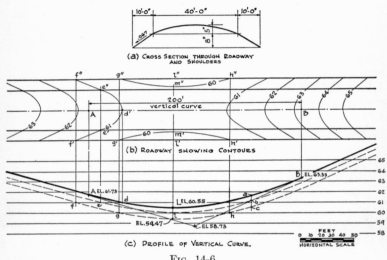

(a) CROSS SECTION THROUGH ROADWAY AND SHOULDERS

(b) ROADWAY SHOWING CONTOURS

(c) PROFILE OF VERTICAL CURVE.

FIG. 14-6.

shoulders of the road. A sufficient number of such points are plotted to enable us to draw the two curved dash lines which represent the shoulder lines in profile.

In Fig. 14-6 (c) it is seen that the horizontal lines that represent the levels of the contours intersect the center line of the roadway and edges of the shoulders at various points. As an example, contour 61 intersects the center line of the roadway (to the left of the low point *L*) at point *d* and the shoulder lines at points *e* and *f*. By projecting these points up to the plan, Fig. 14-6 (b), we establish points *d'*, *e'*, *e''*, *f'*, and *f''*; these are points on contour line 61, and a line connecting them is contour 61. The remaining contours are drawn in a similar manner.

Note that contour level 60 intersects the outer edge of the

shoulder at points g and h. In Art. 14-4 we found the elevation of L, the low point on the center line of the road, to be 60.55. Hence the elevation of point l is $60.55 - 1.08$, or 59.47. Now points g, l, and h are projected up to the plan and we establish points g', l', and h'. The slope of the shoulders is 8″ in 10′, a grade of $0.67/10$, or 0.067. Contour 60 is $(60 - 59.47)$, or 0.53′ above point l. Therefore $l'm'$ and $l''m''$ are each $0.53/0.067$, or 7.9′. Curves drawn through $g'm'h'$ and $g''m''h''$ give us contour 60. The contours crossing the vertical curve, shown in Fig. 14-6 (b), are not necessarily parabolas or straight lines. The method described is the usual procedure for plotting contours on vertical curves. If greater accuracy is desired, additional points on the contours are found by selecting other lines, such as the quarter points of the road and half points on the shoulders, with the corresponding profile curves.

CHAPTER 15

DRAINAGE AND GRADING

15-1. Provision for Drainage. Rain water that falls on the surface of a property either evaporates, percolates into the soil, flows off the site, or drains to some point or points on the site. That portion that does not enter the soil is called the *runoff*, and provision must be made for this excess water. The grading must be so designed that surface water will flow *away from the buildings.* This may necessitate drainage channels with catch basins and a system of underground piping.

When grading is required adjacent to existing trees, the natural elevations should be disturbed as little as possible. The elevation should never be lowered more than 6″. If the ground surface must be raised, stone or brick tree-walls should be provided. They should never exceed 4′ in height.

15-2. Lawns and Seeded Areas. The preferred grade for lawns or seeded slopes adjacent to buildings is 2%; the minimum is 1%. Earth banks should have a 1:2 maximum slope, and, if power lawn mowers are to be used, the slope should not exceed 1:3. Water flowing over banks tends to erode the surface and to wash out planting. To avoid this, drainage gutters may be constructed at the top of the bank to intercept the water.

15-3. Walks and Paths. It is desirable that walks and paths have a crowned cross section so that surface water is diverted to both sides. In many instances this is impossible and small quantities are permitted to flow across the pathway. For these walks, as well as for pavements adjacent to buildings, the cross slope should have a grade of 2% or 3%. To provide for drainage, a 1% longitudinal slope is the preferred minimum. In cold climates, where ice forms readily, 6% is considered to be the maximum longitudinal slope but, in milder climates, the maximum may be as much as 8%. In general, long walks should have minimum

174

slopes but for short walks or short sections of walks the slope may, if necessary, approach the maximum. Steep slopes should always be avoided near the entrances to buildings.

Main entrance walks to residences should have a minimum width of 3′; for public or semi-public buildings a 5′ width is considered to be the minimum. Intersecting paths should have slanted or rounded corners with a minimum radius of 6′.

When possible, steps should be avoided in walks. When they are unavoidable, not less than three risers should be constructed since a smaller number may serve as a stumbling block. If there are five or more risers, handrails should always be provided.

15-4. Roads and Driveways. For drainage, roads and driveways should have a minimum longitudinal slope of 0.5% but a 1% minimum is preferable. A 6% slope is considered to be the maximum, but for short distances it may be as great as 8% or even 10% in mild climates where icy roads present no problem. At road intersections the grade should not exceed 3%.

Concrete and bituminous-surfaced roads are generally parabolic in cross section and should have a crown of 1/4″ for each foot of half-width. For earth roads the height of the crown should be increased to 1/2 inch per foot of half-width. A dished cross section formed by an inverted parabola is sometimes used for driveways. They are recommended only when concrete is used, and the inverted crown should be 1/2 inch per foot of half-width.

The width of a road is determined by the number of lanes of traffic and the parking requirements. Each traffic lane should have a 10′ minimum width. Parallel parking requires a width of 8′, diagonal parking a minimum of 15′ (preferably 18′), and perpendicular parking a minimum of 19′.

When curves occur in driveways on a property, the radius of the curve to the inside edge should not be less than 20′. Streets on which the traffic speed is limited to 20 miles per hour should have curves of at least 100′ radii to the inner edge. Intersecting streets should have rounded corners with a minimum radius of 15′.

To avoid maintenance expense for the property owners, streets in housing developments are commonly dedicated to the city or township upon completion of the project. To insure their acceptance, care should be taken that the specifications and requirements of the local highway department are rigidly adhered to in the construction of such streets.

15-5. **Catch Basins.** The purpose of a catch basin is to intercept storm water before it enters the sewers. Because of the positions of the inlets and outlets, debris and sediment are retained in the catch basin and are prevented from flowing into the sewer line. They must be cleaned out periodically. Figure 15-1

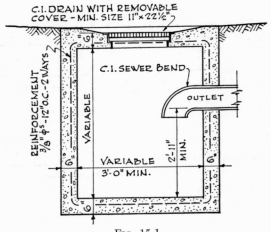

Fig. 15-1.

shows a vertical section taken through a catch basin. The volume of water flowing over a plot should be limited by locating the catch basins so that surface water need not flow more than 75' before entering them. Because of the possibility of their flooding over when stopped up, they should always be placed as far as possible from buildings and walks. They should not be placed closer than 10' to trees or sitting areas.

15-6. **Intensity of Rainfall.** In the design of a drainage system, one of the primary factors is the expected number of inches of rainfall per hour in a given locality. For drainage systems of comparatively small areas, the maximum rainfall in any 2-year period is generally used but, for a more conservative design, the 5-year period may be employed. Data relating to the volume of rainfall may often be obtained from the records kept by municipalities. When such information is not available the two charts shown in Fig. 15-2, prepared by the U. S. Department of Agriculture, may be used. They give the 1-hour rainfall in inches to be expected in the 2- and 5-year periods. One inch of rainfall per hour is equal to approximately 1 ft^3 (actually 1.0083) of water falling on 1 acre of ground per second.

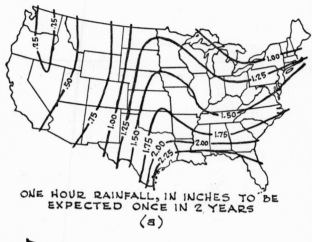

ONE HOUR RAINFALL, IN INCHES TO BE
EXPECTED ONCE IN 2 YEARS
(a)

ONE HOUR RAINFALL, IN INCHES TO BE
EXPECTED ONCE IN 5 YEARS
(b)

From Miscellaneous Publication No. 204 - U. S. Department of Agriculture, by the late David L. Yarnell

FIG. 15-2.

15-7. **Runoff.** All the rain falling on the earth's surface does not reach the drainage lines. Some is lost by evaporation, and some, depending on the porosity of the ground, seeps into the soil. The water that reaches the drainage system is called the *runoff*. Table 7 gives average ratios of runoff to the total amount of rainfall that falls on various surfaces.

The volume of runoff may be estimated by the following formula known as the *rational formula:*

$$Q = ACI$$

in which Q = the runoff from an area, in cubic feet per second.

A = the area to be drained, in acres.

C = the coefficient of runoff. See Table 7.

I = the intensity of rainfall, in inches per hour. If more precise data are not available, use Fig. 15-2.

15-8. **Size of Pipe Required.** When the volume of runoff has been computed, the next step is to determine the proper size of pipe that will carry it away. The quantity of water carried by a pipe depends on several factors among which is the slope of the

TABLE 7. RUNOFF COEFFICIENTS

Roofs	0.95
Concrete or asphalt roads and pavements	0.95
Bituminous macadam roads	0.80
Gravel areas and walks	
loose	0.30
compact	0.70
Vacant lots, unpaved streets	
light plant growth	0.60
no plant growth	0.75
Lawns, parks, golf courses	0.35
Wooded areas	0.20

pipe and the degree of roughness of its interior surface. In formulas for these computations the term n, the friction factor, is generally taken to be 0.015 for vitrified sewer or drainage pipe.

A convenient method of determining the required size of pipe is to use Fig. 15-3, an alignment chart. In this chart the left-hand line gives the volume of runoff and the right-hand line gives various pipe slopes for a friction value of $n = 0.015$. By placing a straightedge connecting the proper values on these two lines

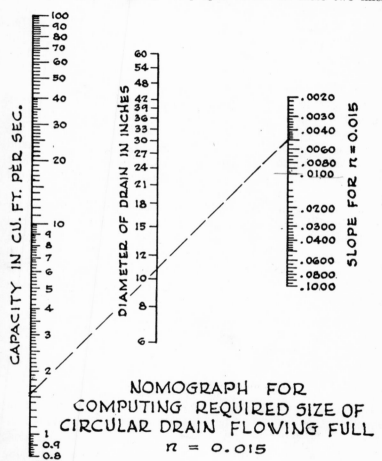

NOMOGRAPH FOR COMPUTING REQUIRED SIZE OF CIRCULAR DRAIN FLOWING FULL
$n = 0.015$

Adapted from the Engineering Manual, War Department, Corps of Engineers, Dec. 1945

FIG. 15-3.

the required pipe size is found on the middle line of the chart. The use of the chart is illustrated in the example given in Art. 15-9.

15-9. Design of Storm Drainage Systems. When the catch basins and roof-drain lines have been located on the site plan, the underground drainage lines are drawn. This may consist of one or more separate systems; they may discharge into a gutter, stream, or public sewer. The drainage system adopted is that which contains the least total length of pipe. Lines should be run with as few bends as possible, using 45° and Y bends at intersections.

The *invert* is the lowest part of the internal surface of a pipe. When the locations of the drainage lines have been established, the elevations of the inverts are computed and marked on the drawing. The pipes should be laid below the frost line and should have a minimum of 3′ of cover to protect them from the weight of passing vehicles. Lines run parallel to the surface of the ground require a minimum of excavation.

Beginning at the upper ends of the lines, the volume of water at the catch basin is determined and the pipe size is established by the use of the chart shown in Fig. 15-3. Water flowing from adjacent properties must be provided for. The drainage line is followed to its point of discharge, accumulating drainage water from other catch basins or branch lines, and the proper pipe sizes established for each additional increase in water volume.

In estimating the volume of water flowing from adjacent properties, consideration must be given to the possibility that vacant land may be built up at a later date.

EXAMPLE

EXAMPLE. A 200′x300′ plot, located in the southern part of Ohio, contains 4,360ft² of buildings, 6,500ft² of bituminous macadam roads, 2,200ft² of concrete walks, and the remainder, 46,940ft², is a seeded and planted area. An adjacent plot, having an area of ⅓ acre, is vacant land the runoff from which will flow onto the 200′x300′ plot. Determine the size of the drainage pipe to accommodate the runoff from this plot if the pipe has a slope of 0.005.

SOLUTION. The first step is to compute the volume of runoff; therefore, we will use the formula $Q = ACI$ given in Art. 15-7 and the C coefficients for runoff given in Table 7. Then,

	Areas, ft^2		Coefficient		Adjusted Area, ft^2
Roofs	4,360	$\times$	0.95	$=$	4,150
Macadam roads	6,500	$\times$	0.80	$=$	5,200
Concrete walks	2,200	$\times$	0.95	$=$	2,100
Lawns	46,940	$\times$	0.35	$=$	16,430
	60,000ft^2				27,880ft^2

Since there are 43,560ft^2 in an acre, $27,880 \div 43,560 = 0.64$ acre, the value of $A \times C$ in the formula exclusive of the adjoining ⅓-acre vacant plot. This land may be built up at a later date; consequently the C coefficient will be taken as 0.75. Then, $0.33 \times 0.75 = 0.25$ and $0.64 + 0.25 = 0.89$, the value of $A \times C$.

The intensity of rainfall, I in the formula, is found in Fig. 15-2 (b) to be 1.75″, the 1-hour rainfall expected once in 5 years. Then $Q = ACI$, or $Q = 0.89 \times 1.75 = 1.56$ft^3 per sec, the volume of runoff from the area to be drained.

Now refer to Fig. 15-3. Place a straightedge so that it connects 1.56 on the left-hand line with 0.005 on the right-hand line as shown on the chart. The straightedge now crosses the middle line between 10″ and 12″. Therefore, we will accept a 12″ diameter drainage pipe.

15-10. **Sewer and Drainage Pipe.** Because of its comparatively low cost, small sewer and drainage lines are usually constructed of vitrified-clay pipe. Inside buildings, and within 10′ of them, only cast-iron pipe should be used. When drainage pipes exceed 15″ in diameter, concrete and asbestos cement pipe are often used. Owing to the fact that roots frequently clog clay pipes, 6″ is considered to be the minimum size although some authorities recommend not less than 8″. For house drains, cast-iron pipe 4″ in diameter is commonly used.

15-11. **Parking Areas.** With the increasing number of problems that result from traffic congestion it has become desirable to provide parking areas on the building site of all but the smallest buildings. These areas are for the accommodation of the occupants as well as for visitors to the buildings. Many municipalities require on-site parking areas for certain types of buildings. In arranging the parking spaces they should be so arranged that parked cars will not obscure the vision at street intersections or curves.

Care should be taken to see that parking areas are properly drained, the minimum and maximum slopes being 0.5% and 4%, respectively. Surface water should not be permitted to flow directly onto public highways but should be intercepted by catch basins or inlets.

CHAPTER 16

STAKING OUT BUILDINGS AND ROADS

16-1. **Laying Off Angles.** The preferred surveying instrument to use in staking out buildings is one in which the telescope may be moved up and down in a vertical plane. This is found in the builders' transit level and the engineers' transit. If the builders' level is used, the plumb bob is employed for transferring levels to points on the ground

To stake out a building a surveying instrument is required to lay off given angles. Suppose that we are given the line AB, Fig.

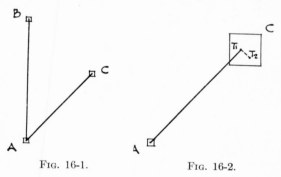

FIG. 16-1.   FIG. 16-2.

16-1, and that we are required to locate point C, 125′ from A, angle BAC being 42° 30′. To begin, the instrument is set up over point A and the zero mark on the horizontal circle is sighted on point B, as described in Art. 4-11. Next, turn the instrument clockwise 42° 30′ and at 125′ from A drive a stake. By directions from the instrument man a tack is driven in the stake marking the location of point C.

Angle BAC is now "measured by repetition," as explained in Art. 4-11. Suppose that we now find the angle to be 42° 28′ 30″. This shows that there was an error of 0° 1′ 30″ in setting the

183

tack; it should have been placed a certain distance farther to the right. Now let us compute this distance. By referring to a table of natural tangents, we find that the tangent of 1′ is 0.0003. Therefore, for each minute of error multiply the length AC by 0.0003 and for each second of error multiply the length AC by 0.000005. Hence, for this problem in which the error is $0° 1′ 30″$,

$$1 \times 0.0003 = 0.00030$$
$$30 \times 0.000005 = 0.00015$$
$$\overline{ 0.00045}$$

Since AC is 125′, the total error is 0.00045×125, or 0.056′.

Figure 16-2 shows the stake driven at point C, and T_1 indicates the position of the first tack. Since we have found that the exact position of point C is 0.056′ farther to the right, we measure this distance (on a line at right angles with AC) and drive a second tack, T_2, thus establishing the corrected position of point C. The first tack should now be removed. For problems like this, in which the angle for the first tack was found to be in error, care must be taken to measure off in the proper direction the distance needed to correct the error.

16-2. Staking Out Buildings. The boundaries of building plots should be established by monuments (markers), at the corners, set in place by a registered surveyor. The site plan shows the positions of the boundaries and their distances from the proposed

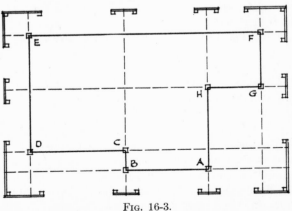

FIG. 16-3.

buildings. The builder locates the boundaries of the plot, and from the data shown on the plot plan lays out the building by placing stakes at the corners. This will result in one or more rectangles, as shown by the area *ABCDEFGH* in Fig. 16-3. To check the accuracy of the work, the two diagonals of the rectangles can be measured. If there is no error, the diagonals will be equal in length.

16-3. Batter Boards. The stakes locating the corners of the building, referred to in Art. 16-2, will be displaced during the

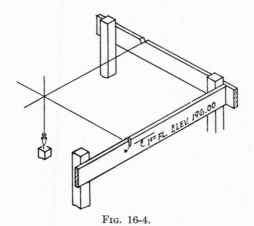

FIG. 16-4.

excavation operations. In order that their locations may be retained, *batter boards* are used. They are placed at sufficient distances (generally 5′ or 10′) outside the perimeter of the structure so that they will not be disturbed by the building operations.

Batter boards consist of vertical stakes (2″x4″′s or 4″x4″′s) driven firmly into the ground with 1″ or 2″ boards nailed across them, as shown in Fig. 16-4. Notches or saw cuts are cut in the boards so that cords or wires may be strung from one batter board to another. The intersections of the cords are directly above the corners of the building. The batter boards should be at the same height, and this height should coincide with a floor level or bear some definite relation to it. The elevation should be plainly marked on the batter boards as indicated in the figure. The elevations of the batter boards should be checked frequently during the excavation and foundation work to see that they have not shifted or been displaced.

16-4. Setting Batter Boards. After the building has been staked out with the corner stakes, the stakes for the batter boards are driven. The surveying instrument is now set up and levelled in a position so that sights may be taken on the bench mark and the batter boards. Suppose, for example, the bench mark is at elevation 197.53 and that the first-floor elevation is to be 196.00. A backsight is taken on the bench mark, and the reading is found to be 4.68. Then, since the bench mark is 197.53, 197.53 + 4.68

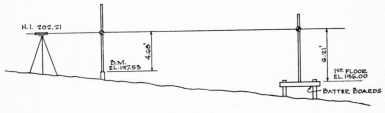

Fig. 16-5.

= 202.21, the height of the instrument, H.I. The difference between H.I. and the first-floor elevation is 202.21 − 196.00, or 6.21'. These heights and elevations are shown in Fig. 16-5. The target on the rod is set at 6.21', and the rod is placed alongside one of the batter-board stakes. The rod is moved up and down until the horizontal cross hair coincides with the target. The height of the bottom of the rod is then marked on the stake, and batter boards are nailed to the stakes so that their tops are level with this mark. The carpenters' level is used to level the batter boards. This procedure is repeated for each set of batter boards.

The instrument is now set up over one of the stakes marking a corner of the building, and the lines of the building are projected to locate the saw cuts and notches on the batter boards. For example, the instrument is set up over point A, Fig. 16-3, and sighted on point B. This establishes line of sight AB, and a saw cut is placed on the batter board. A less accurate but more rapid method would be to stand to the right of point A and to sight-in the point on the batter board by eye. By this method points may be located with an error generally not greater than $\frac{1}{8}''$.

16-5. **Laying Out Column Footings.** When the excavations for the building have extended to the lowest level, the centers of the column footings are located and stakes are driven at the center points. The marks on the batter boards may be used in locating the column centers. If the column footing is large it may require its own batter boards to establish its perimeter. For small footings, stakes are driven in the shape of a rectangle a whole number of feet (2' or 3' depending on conditions) outside the footing perimeter. For convenience in excavating the footing to the proper depth, these stakes are driven so that their tops are at the level of the top of the footing. Lines stretched from stake to stake may be used to locate the centers of the column footings, and lines are also used to establish the perimeter of the footing.

16-6. **Staking Out Roads and Driveways.** In laying out roads and driveways the usual procedure is to drive stakes along the center lines. These stakes are located at each point of change of slope and at regular intervals, 25' or 50', along the length of the drive. The elevations of the various points are taken from the site plan. The stakes are driven so that their tops are at the elevation of the finished driveway or at a number of full feet above the required elevation and are so marked. They are first driven with their tops several inches above the desired elevation. The rod, with the target placed at the required setting, is held on top of the stake, and the stake is then driven into the ground until the center of the target coincides with the level line of sight of the instrument. On straight runs of the road, stakes may be located on "offset lines" a number of full feet away from and parallel to the center line of the road. These stakes are so located that they will not be disturbed during the construction operations. The methods used for locating points on circular and vertical curves are given in Chapters 9 and 14, respectively.

16-7. **Grades of Constant Slope.** When a number of intermediate stakes are to be set on a portion of a road that has a constant slope, the following method may be used: Stakes are first located and driven to the required elevations at the two ends of the slope. The instrument is now set up over one of these stakes, and the height from the stake to the center line of the telescope is measured. The target on the rod is then set at this measured height, and the rod is held over the stake at the other end of the constant slope. Now the telescope is tilted so that the horizontal cross hair coincides with the target. The line of sight of the instrument is now parallel with the slope of the road. Without altering the position of the target, the rod is then held over the successive intermediate stakes which are now driven to a depth that permits the center of the target to coincide with the line of sight of the instrument.

16-8. Setting Stakes for Grading. When an area of a plot is to be graded to establish new elevations for the surface of the ground, an adequate number of stakes are driven to aid in establishing the new elevations. It is customary to drive these stakes so that their tops are a full number of feet above or below the finished elevations. After they are driven they should be plainly marked, such as "4' fill" or "2' cut."

16-9. Staking Out Sewers. In staking out a sewer line stakes are first driven on its center line at regular intervals of 25' to 50'. On each side of this line of stakes, far enough apart to prevent them from being disturbed by the trench excavation, heavy stakes are driven to which planks are nailed across the sewer line. Small boards, placed vertically, are nailed to the planks with one edge in alignment with the center line of the sewer. A nail is now partially driven into the vertical board at some full number of feet above the *invert* of the sewer pipe. The invert is the lowest part of the internal surface of the pipe. Cords or wires stretched from nail to nail mark a line at some established height above the invert, and a vertical board held beside the string is used to lay the sewer pipe at the desired elevation. The drawings showing the sewer lines should include a profile of the sewer, showing its elevations at various points and their relation to the surface of the ground.

CHAPTER 17

SITE SELECTION CONSIDERATIONS

17-1. Site Selection Considerations. The items enumerated in this chapter include the most important factors that should be considered when purchasing a site for building purposes; they particularly relate to sites used for residential buildings. Of the suggested items, many are not pertinent for certain locations. It is not the purpose to comment on the various items; most of them are merely mentioned, but their consideration will be helpful in evaluating the advantages and disadvantages of two or more prospective sites. Certain facts concerning a location may be difficult to obtain, and consultation with property owners in the neighborhood is invariably a source of helpful information.

17-2. Preliminary Considerations

1. Examine survey or deed description showing dimensions and shape of the lot.
2. Location, availability, and cost of public services and utilities.
    (*a*) Water supply, water pressure.
    (*b*) Storm and sanitary sewers.
    (*c*) Electricity, underground or overhead.
    (*d*) Gas.
    (*e*) Telephone, underground or overhead.
    (*f*) Mail delivery.
    (*g*) Garbage disposal, incinerators, etc.
    (*h*) Fire and police protection.
    (*i*) Tax rate and assessments.
    (*j*) Condition of roads.
    (*k*) Snow removal.
3. Deed restrictions. This may require an examination of previous deeds.
    (*a*) Limitation of size and type of building, materials, and cost.

189

(b) Easements, rights of way.

(c) Party wall and fence considerations.

(d) Restrictions on use of pre-fabricated buildings.

4. Local and state codes and ordinances.

(a) Zoning, land coverage, set-back lines, location of trees. Absence of good zoning ordinances is detrimental.

(b) Building codes, fire protection requirements, exits.

(c) Health and sanitation requirements, plumbing codes.

(d) Permits.

17-3. Environmental Factors

1. Approaches. Are main approaches through desirable neighborhoods?

2. Character of neighborhood. Is it in the process of change? Building activity. Types and cost of buildings. Desirability of neighbors, age and income brackets. Is population increasing? Consult city and regional planning commission or chamber of commerce.

3. Transportation

(a) Types, frequency, and cost.

(b) Total travel time to place of employment, schools, shopping centers, cultural and recreational facilities.

4. Shopping facilities.

5. Schools.

6. Churches.

7. Libraries, museums, places of amusements.

8. Recreation facilities, parks, and swimming pools.

9. Medical and hospital facilities.

17-4. Hazards to Life

1. Highway traffic. High-speed traffic routes and intersections. Curved streets which do not permit through traffic reduce speed and tend to avoid accidents.

2. Railway grade crossings.

3. Flood areas. Low sites adjacent to streams or rivers may flood over.

4. Proximity to storage tanks for oil, gas, or other inflammable material. Industrial hazards.

5. Proximity to airports. The danger of airplane crashes.

17-5. Nuisances

1. Objectionable noises from highway traffic especially at intersections and on hills, railways, industrial operations, airfields, parking lots, schools, and playgrounds. Baffles of planting or other material are frequently employed to reduce objectionable noises.

2. Smoke from railroads, incinerators, and factories.

3. Dust from vacant lots and untreated dirt playgrounds.

4. Offensive odors from highway traffic, incinerators, polluted streams, factories, and farm animals.

17-6. Important Features of Sites

1. SIZE. Is the size adequate for the proposed buildings with driveways, garage, and future expansion? Is there sufficient area for parking?

2. SHAPE. Will the shape of the plot permit all the buildings and required facilities? Thoroughly investigate lots that are long and narrow or triangular.

3. TOPOGRAPHY. From the standpoint of economy, gently sloping sites with good drainage are preferable to very flat plots or very steep slopes. Flat slopes may require excessive grading or deep sewer lines. Does the site permit a good location for sewage disposal? Steep slopes, unless well planted, are subject to erosion. Buildings at the foot of slopes may be flooded at certain seasons. Investigate possibility of water in basements. If drainage from basement is required, where will it discharge?

4. FILLED GROUND. Foundations on filled ground are subject to unequal settlement. Check the depth and age of the fill.

5. ROCK FOUNDATION BEDS. Rock foundation beds provide minimum settlement, but the cost of excavation may be prohibitive. Consult local builders concerning underground conditions. Test pits or borings may be required to determine subsoil conditions.

6. TOPSOIL. Existing topsoil should be examined. If it lacks the qualities that will provide good growing conditions it may require reconditioning. Unsuitable topsoil may demand that new material be brought to the site. Consult your landscape architect.

7. ORIENTATION. In planning the building consideration of the orientation is of the utmost importance. Always consider the

effect of direct sunlight in the important rooms of the building,
the direction of the prevailing winds, and the desirable views.

8. PRIVACY.   If privacy is desired by the owner this must be
borne in mind in both planning the building as well as its location
on the site.

9. ADJACENT PROPERTIES.   Corner lots in built-up neighborhoods may have the advantage of providing light and air but the
disadvantages of less privacy, traffic noises, increased cost of land,
sidewalks, and utilities.   "Key lots," as indicated in Fig. 17-1,

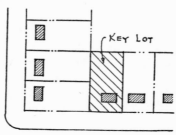

FIG. 17-1.

because of the positions of the adjacent properties, are objectionable.   The better types of development avoid this condition.
Building sites adjacent to public driveways are undesirable.   When
plots front on private streets these streets are not generally maintained by the city or township, and the cost of maintenance must
be assumed by the owner.

10. LANDSCAPING.   If the building project requires landscaping,
a landscape architect should be consulted when the building is
being designed.   He is frequently brought in too late.   Landscaping is a vital part of the building project.   In budgeting the
cost of the operation, ample provision should be made for this
work.   Usually, well-planned landscaping enhances the value of
the property far beyond the cost of such work.

11. BEST USE OF SITE.   The prime consideration in any site
is whether or not the site is being developed according to its
highest potentialities.   If another type of building would seem to
be more appropriate for the site it is doubtful whether the con-
templated building would ever be completely successful.

CHAPTER 18

CHECK LIST FOR SITE PLANS

18-1. Check List. The following list is an enumeration of some of the most important items that may be required on site plans. It should prove to be helpful in preventing the omission of important data.

1. Complete title, location, owner, architect or engineer, and date. List of symbols and abbreviations.

2. Scale of drawing.

3. Direction of true north.

4. Complete dimensions of building and property lines including angles and all dimensions required for locating the building on the site. Positions of drives, paths, and existing buildings. If curves are shown, give radii, P.T.'s, P.C.'s, length of curves, and included angles. All dimensions should be given in *U. S. Standard.*

5. Zoning setbacks and restriction lines.

6. Area of plot.

7. Show limit of contract lines.

8. Details and location of curbs and their radii, dropped curbs, and tree enclosures.

9. Datum, existing contours, and contours for proposed work.

10. Basement and first-floor elevations.

11. Adjoining properties and buildings with names of owners.

12. Indicate buildings, old walls, paving, and curbs to be removed.

13. Show location, ground elevation, name and size of trees and shrubbery to be removed or to remain.

14. Show elevations and slopes of streets, driveways, paths, and parking areas with elevations of high and low points.

15. Show sections through driveways.

16. Show yard drains, catch basins, and curb inlets, and give elevations and sizes.

17. Show locations and data relating to utilities, sewers, and sewage-disposal systems.

18. Show existing and proposed manholes, telephone poles, or conduits, fire hydrants, and light standards.

19. Show sections through and give top and bottom elevations of walls.

20. Show all steps, ramps, railings, areaways, and gratings.

21. Show locations and extent of surfacing materials, lawns, and planted or sodded areas.

22. If planting is required, show planting schedule, giving sizes of plants and trees with both botanical and common names. The number of individual trees and plants should be shown on the plan.

LOGARITHMIC
TABLES

TABLE 1. LOGARITHMS OF NUMBERS

N	0	1	2	3	4	5	6	7	8	9
100	00000	00043	00087	00130	00173	00217	00260	00303	00346	00389
1	0432	0475	0518	0561	0604	0647	0689	0732	0775	0817
2	0860	0903	0945	0988	1030	1072	1115	1157	1199	1242
3	1284	1326	1368	1410	1452	1494	1536	1578	1620	1662
4	1703	1745	1787	1828	1870	1912	1953	1995	2036	2078
5	2119	2160	2202	2243	2284	2325	2366	2407	2449	2490
6	2531	2572	2612	2653	2694	2735	2776	2816	2857	2898
7	2938	2979	3019	3060	3100	3141	3181	3222	3262	3302
8	3342	3383	3423	3463	3503	3543	3583	3623	3663	3703
9	3743	3782	3822	3862	3902	3941	3981	4021	4060	4100
110	04139	04179	04218	04258	04297	04336	04376	04415	04454	04493
1	4532	4571	4610	4650	4689	4727	4766	4805	4844	4883
2	4922	4961	4999	5038	5077	5115	5154	5192	5231	5269
3	5308	5346	5385	5423	5461	5500	5538	5576	5614	5652
4	5690	5729	5767	5805	5843	5881	5918	5956	5994	6032
5	6070	6108	6145	6183	6221	6258	6296	6333	6371	6408
6	6446	6483	6521	6558	6595	6633	6670	6707	6744	6781
7	6819	6856	6893	6930	6967	7004	7041	7078	7115	7151
8	7188	7225	7262	7298	7335	7372	7408	7445	7482	7518
9	7555	7591	7628	7664	7700	7737	7773	7809	7846	7882
120	07918	07954	07990	08027	08063	08099	08135	08171	08207	08243
1	8279	8314	8350	8386	8422	8458	8493	8529	8565	8600
2	8636	8672	8707	8743	8778	8814	8849	8884	8920	8955
3	8991	9026	9061	9096	9132	9167	9202	9237	9272	9307
4	9342	9377	9412	9447	9482	9517	9552	9587	9621	9656
5	9691	9726	9760	9795	9830	9864	9899	9934	9968	10003
6	10037	10072	10106	10140	10175	10209	10243	10278	10312	0346
7	0380	0415	0449	0483	0517	0551	0585	0619	0653	0687
8	0721	0755	0789	0823	0857	0890	0924	0958	0992	1025
9	1059	1093	1126	1160	1193	1227	1261	1294	1327	1361
130	11394	11428	11461	11494	11528	11561	11594	11628	11661	11694
1	1727	1760	1793	1826	1860	1893	1926	1959	1992	2024
2	2057	2090	2123	2156	2189	2222	2254	2287	2320	2352
3	2385	2418	2450	2483	2516	2548	2581	2613	2646	2678
4	2710	2743	2775	2808	2840	2872	2905	2937	2969	3001
5	3033	3066	3098	3130	3162	3194	3226	3258	3290	3322
6	3354	3386	3418	3450	3481	3513	3545	3577	3609	3640
7	3672	3704	3735	3767	3799	3830	3862	3893	3925	3956
8	3988	4019	4051	4082	4114	4145	4176	4208	4239	4270
9	4301	4333	4364	4395	4426	4457	4489	4520	4551	4582
140	14613	14644	14675	14706	14737	14768	14799	14829	14860	14891
1	4922	4953	4983	5014	5045	5076	5106	5137	5168	5198
2	5229	5259	5290	5320	5351	5381	5412	5442	5473	5503
3	5534	5564	5594	5625	5655	5685	5715	5746	5776	5806
4	5836	5866	5897	5927	5957	5987	6017	6047	6077	6107
5	6137	6167	6197	6227	6256	6286	6316	6346	6376	6406
6	6435	6465	6495	6524	6554	6584	6613	6643	6673	6702
7	6732	6761	6791	6820	6850	6879	6909	6938	6967	6997
8	7026	7056	7085	7114	7143	7173	7202	7231	7260	7289
9	7319	7348	7377	7406	7435	7464	7493	7522	7551	7580
150	17609	17638	17667	17696	17725	17754	17782	17811	17840	17869

TABLE 1. LOGARITHMS OF NUMBERS

N	0	1	2	3	4	5	6	7	8	9
150	17609	17638	17667	17696	17725	17754	17782	17811	17840	17869
1	7898	7926	7955	7984	8013	8041	8070	8099	8127	8156
2	8184	8213	8241	8270	8298	8327	8355	8384	8412	8441
3	8469	8498	8526	8554	8583	8611	8639	8667	8696	8724
4	8752	8780	8808	8837	8865	8893	8921	8949	8977	9005
5	9033	9061	9089	9117	9145	9173	9201	9229	9257	9285
6	9312	9340	9368	9396	9424	9451	9479	9507	9535	9562
7	9590	9618	9645	9673	9700	9728	9756	9783	9811	9838
8	9866	9893	9921	9948	9976	20003	20030	20058	20085	20112
9	20140	20167	20194	20222	20249	0276	0303	0330	0358	0385
160	20412	20439	20466	20493	20520	20548	20575	20602	20629	20656
1	0683	0710	0737	0763	0790	0817	0844	0871	0898	0925
2	0952	0978	1005	1032	1059	1085	1112	1139	1165	1192
3	1219	1245	1272	1299	1325	1352	1378	1405	1431	1458
4	1484	1511	1537	1564	1590	1617	1643	1669	1696	1722
5	1748	1775	1801	1827	1854	1880	1906	1932	1958	1985
6	2011	2037	2063	2089	2115	2141	2167	2194	2220	2246
7	2272	2298	2324	2350	2376	2401	2427	2453	2479	2505
8	2531	2557	2583	2608	2634	2660	2686	2712	2737	2763
9	2789	2814	2840	2866	2891	2917	2943	2968	2994	3019
170	23045	23070	23096	23121	23147	23172	23198	23223	23249	23274
1	3300	3325	3350	3376	3401	3426	3452	3477	3502	3528
2	3553	3578	3603	3629	3654	3679	3704	3729	3754	3779
3	3805	3830	3855	3880	3905	3930	3955	3980	4005	4030
4	4055	4080	4105	4130	4155	4180	4204	4229	4254	4279
5	4304	4329	4353	4378	4403	4428	4452	4477	4502	4527
6	4551	4576	4601	4625	4650	4674	4699	4724	4748	4773
7	4797	4822	4846	4871	4895	4920	4944	4969	4993	5018
8	5042	5066	5091	5115	5139	5164	5188	5212	5237	5261
9	5285	5310	5334	5358	5382	5406	5431	5455	5479	5503
180	25527	25551	25575	25600	25624	25648	25672	25696	25720	25744
1	5768	5792	5816	5840	5864	5888	5912	5935	5959	5983
2	6007	6031	6055	6079	6102	6126	6150	6174	6198	6221
3	6245	6269	6293	6316	6340	6364	6387	6411	6435	6458
4	6482	6505	6529	6553	6576	6600	6623	6647	6670	6694
5	6717	6741	6764	6788	6811	6834	6858	6881	6905	6928
6	6951	6975	6998	7021	7045	7068	7091	7114	7138	7161
7	7184	7207	7231	7254	7277	7300	7323	7346	7370	7393
8	7416	7439	7462	7485	7508	7531	7554	7577	7600	7623
9	7646	7669	7692	7715	7738	7761	7784	7807	7830	7852
190	27875	27898	27921	27944	27967	27989	28012	28035	28058	28081
1	8103	8126	8149	8171	8194	8217	8240	8262	8285	8307
2	8330	8353	8375	8398	8421	8443	8466	8488	8511	8533
3	8556	8578	8601	8623	8646	8668	8691	8713	8735	8758
4	8780	8803	8825	8847	8870	8892	8914	8937	8959	8981
5	9003	9026	9048	9070	9092	9115	9137	9159	9181	9203
6	9226	9248	9270	9292	9314	9336	9358	9380	9403	9425
7	9447	9469	9491	9513	9535	9557	9579	9601	9623	9645
8	9667	9688	9710	9732	9754	9776	9798	9820	9842	9863
9	9885	9907	9929	9951	9973	9994	30016	30038	30060	30081
200	30103	30125	30146	30168	30190	30211	30233	30255	30276	30298

TABLE 1. LOGARITHMS OF NUMBERS

N	0	1	2	3	4	5	6	7	8	9
200	30103	30125	30146	30168	30190	30211	30233	30255	30276	30298
1	0320	0341	0363	0384	0406	0428	0449	0471	0492	0514
2	0535	0557	0578	0600	0621	0643	0664	0685	0707	0728
3	0750	0771	0792	0814	0835	0856	0878	0899	0920	0942
4	0963	0984	1006	1027	1048	1069	1091	1112	1133	1154
5	1175	1197	1218	1239	1260	1281	1302	1323	1345	1366
6	1387	1408	1429	1450	1471	1492	1513	1534	1555	1576
7	1597	1618	1639	1660	1681	1702	1723	1744	1765	1785
8	1806	1827	1848	1869	1890	1911	1931	1952	1973	1994
9	2015	2035	2056	2077	2098	2118	2139	2160	2181	2201
210	32222	32243	32263	32284	32305	32325	32346	32366	32387	32408
1	2428	2449	2469	2490	2510	2531	2552	2572	2593	2613
2	2634	2654	2675	2695	2715	2736	2756	2777	2797	2818
3	2838	2858	2879	2899	2919	2940	2960	2980	3001	3021
4	3041	3062	3082	3102	3122	3143	3163	3183	3203	3224
5	3244	3264	3284	3304	3325	3345	3365	3385	3405	3425
6	3445	3465	3486	3506	3526	3546	3566	3586	3606	3626
7	3646	3666	3686	3706	3726	3746	3766	3786	3806	3826
8	3846	3866	3885	3905	3925	3945	3965	3985	4005	4025
9	4044	4064	4084	4104	4124	4143	4163	4183	4203	4223
220	34242	34262	34282	34301	34321	34341	34361	34380	34400	34420
1	4439	4459	4479	4498	4518	4537	4557	4577	4596	4616
2	4635	4655	4674	4694	4713	4733	4753	4772	4792	4811
3	4830	4850	4869	4889	4908	4928	4947	4967	4986	5005
4	5025	5044	5064	5083	5102	5122	5141	5160	5180	5199
5	5218	5238	5257	5276	5295	5315	5334	5353	5372	5392
6	5411	5430	5449	5468	5488	5507	5526	5545	5564	5583
7	5603	5622	5641	5660	5679	5698	5717	5736	5755	5774
8	5793	5813	5832	5851	5870	5889	5908	5927	5946	5965
9	5984	6003	6021	6040	6059	6078	6097	6116	6135	6154
230	36173	36192	36211	36229	36248	36267	36286	36305	36324	36342
1	6361	6380	6399	6418	6436	6455	6474	6493	6511	6530
2	6549	6568	6586	6605	6624	6642	6661	6680	6698	6717
3	6736	6754	6773	6791	6810	6829	6847	6866	6884	6903
4	6922	6940	6959	6977	6996	7014	7033	7051	7070	7088
5	7107	7125	7144	7162	7181	7199	7218	7236	7254	7273
6	7291	7310	7328	7346	7365	7383	7401	7420	7438	7457
7	7475	7493	7511	7530	7548	7566	7585	7603	7621	7639
8	7658	7676	7694	7712	7731	7749	7767	7785	7803	7822
9	7840	7858	7876	7894	7912	7931	7949	7967	7985	8003
240	38021	38039	38057	38075	38093	38112	38130	38148	38166	38184
1	8202	8220	8238	8256	8274	8292	8310	8328	8346	8364
2	8382	8399	8417	8435	8453	8471	8489	8507	8525	8543
3	8561	8578	8596	8614	8632	8650	8668	8686	8703	8721
4	8739	8757	8775	8792	8810	8828	8846	8863	8881	8899
5	8917	8934	8952	8970	8987	9005	9023	9041	9058	9076
6	9094	9111	9129	9146	9164	9182	9199	9217	9235	9252
7	9270	9287	9305	9322	9340	9358	9375	9393	9410	9428
8	9445	9463	9480	9498	9515	9533	9550	9568	9585	9602
9	9620	9637	9655	9672	9690	9707	9724	9742	9759	9777
250	39794	39811	39829	39846	39863	39881	39898	39915	39933	39950

TABLE 1. LOGARITHMS OF NUMBERS

N	0	1	2	3	4	5	6	7	8	9
250	39794	39811	39829	39846	39863	39881	39898	39915	39933	39950
1	9967	9985	40002	40019	40037	40054	40071	40088	40106	40123
2	40140	40157	0175	0192	0209	0226	0243	0261	0278	0295
3	0312	0329	0346	0364	0381	0398	0415	0432	0449	0466
4	0483	0500	0518	0535	0552	0569	0586	0603	0620	0637
5	0654	0671	0688	0705	0722	0739	0756	0773	0790	0807
6	0824	0841	0858	0875	0892	0909	0926	0943	0960	0976
7	0993	1010	1027	1044	1061	1078	1095	1111	1128	1145
8	1162	1179	1196	1212	1229	1246	1263	1280	1296	1313
9	1330	1347	1363	1380	1397	1414	1430	1447	1464	1481
260	41497	41514	41531	41547	41564	41581	41597	41614	41631	41647
1	1664	1681	1697	1714	1731	1747	1764	1780	1797	1814
2	1830	1847	1863	1880	1896	1913	1929	1946	1963	1979
3	1996	2012	2029	2045	2062	2078	2095	2111	2127	2144
4	2160	2177	2193	2210	2226	2243	2259	2275	2292	2308
5	2325	2341	2357	2374	2390	2406	2423	2439	2455	2472
6	2488	2504	2521	2537	2553	2570	2586	2602	2619	2635
7	2651	2667	2684	2700	2716	2732	2749	2765	2781	2797
8	2813	2830	2846	2862	2878	2894	2911	2927	2943	2959
9	2975	2991	3008	3024	3040	3056	3072	3088	3104	3120
270	43136	43152	43169	43185	43201	43217	43233	43249	43265	43281
1	3297	3313	3329	3345	3361	3377	3393	3409	3425	3441
2	3457	3473	3489	3505	3521	3537	3553	3569	3584	3600
3	3616	3632	3648	3664	3680	3696	3712	3727	3743	3759
4	3775	3791	3807	3823	3838	3854	3870	3886	3902	3917
5	3933	3949	3965	3981	3996	4012	4028	4044	4059	4075
6	4091	4107	4122	4138	4154	4170	4185	4201	4217	4232
7	4248	4264	4279	4295	4311	4326	4342	4358	4373	4389
8	4404	4420	4436	4451	4467	4483	4498	4514	4529	4545
9	4560	4576	4592	4607	4623	4638	4654	4669	4685	4700
280	44716	44731	44747	44762	44778	44793	44809	44824	44840	44855
1	4871	4886	4902	4917	4932	4948	4963	4979	4994	5010
2	5025	5040	5056	5071	5086	5102	5117	5133	5148	5163
3	5179	5194	5209	5225	5240	5255	5271	5286	5301	5317
4	5332	5347	5362	5378	5393	5408	5423	5439	5454	5469
5	5484	5500	5515	5530	5545	5561	5576	5591	5606	5621
6	5637	5652	5667	5682	5697	5712	5728	5743	5758	5773
7	5788	5803	5818	5834	5849	5864	5879	5894	5909	5924
8	5939	5954	5969	5984	6000	6015	6030	6045	6060	6075
9	6090	6105	6120	6135	6150	6165	6180	6195	6210	6225
290	46240	46255	46270	46285	46300	46315	46330	46345	46359	46374
1	6389	6404	6419	6434	6449	6464	6479	6494	6509	6523
2	6538	6553	6568	6583	6598	6613	6627	6642	6657	6672
3	6687	6702	6716	6731	6746	6761	6776	6790	6805	6820
4	6835	6850	6864	6879	6894	6909	6923	6938	6953	6967
5	6982	6997	7012	7026	7041	7056	7070	7085	7100	7114
6	7129	7144	7159	7173	7188	7202	7217	7232	7246	7261
7	7276	7290	7305	7319	7334	7349	7363	7378	7392	7407
8	7422	7436	7451	7465	7480	7494	7509	7524	7538	7553
9	7567	7582	7596	7611	7625	7640	7654	7669	7683	7698
300	47712	47727	47741	47756	47770	47784	47799	47813	47828	47842

TABLE 1. LOGARITHMS OF NUMBERS

N	0	1	2	3	4	5	6	7	8	9
300	47712	47727	47741	47756	47770	47784	47799	47813	47828	47842
1	7857	7871	7885	7900	7914	7929	7943	7958	7972	7986
2	8001	8015	8029	8044	8058	8073	8087	8101	8116	8130
3	8144	8159	8173	8187	8202	8216	8230	8244	8259	8273
4	8287	8302	8316	8330	8344	8359	8373	8387	8401	8416
5	8430	8444	8458	8473	8487	8501	8515	8530	8544	8558
6	8572	8586	8601	8615	8629	8643	8657	8671	8686	8700
7	8714	8728	8742	8756	8770	8785	8799	8813	8827	8841
8	8855	8869	8883	8897	8911	8926	8940	8954	8968	8982
9	8996	9010	9024	9038	9052	9066	9080	9094	9108	9122
310	49136	49150	49164	49178	49192	49206	49220	49234	49248	49262
1	9276	9290	9304	9318	9332	9346	9360	9374	9388	9402
2	9415	9429	9443	9457	9471	9485	9499	9513	9527	9541
3	9554	9568	9582	9596	9610	9624	9638	9651	9665	9679
4	9693	9707	9721	9734	9748	9762	9776	9790	9803	9817
5	9831	9845	9859	9872	9886	9900	9914	9927	9941	9955
6	9969	9982	9996	50010	50024	50037	50051	50065	50079	50092
7	50106	50120	50133	0147	0161	0174	0188	0202	0215	0229
8	0243	0256	0270	0284	0297	0311	0325	0338	0352	0365
9	0379	0393	0406	0420	0433	0447	0461	0474	0488	0501
320	50515	50529	50542	50556	50569	50583	50596	50610	50623	50637
1	0651	0664	0678	0691	0705	0718	0732	0745	0759	0772
2	0786	0799	0813	0826	0840	0853	0866	0880	0893	0907
3	0920	0934	0947	0961	0974	0987	1001	1014	1028	1041
4	1055	1068	1081	1095	1108	1121	1135	1148	1162	1175
5	1188	1202	1215	1228	1242	1255	1268	1282	1295	1308
6	1322	1335	1348	1362	1375	1388	1402	1415	1428	1441
7	1455	1468	1481	1495	1508	1521	1534	1548	1561	1574
8	1587	1601	1614	1627	1640	1654	1667	1680	1693	1706
9	1720	1733	1746	1759	1772	1786	1799	1812	1825	1838
330	51851	51865	51878	51891	51904	51917	51930	51943	51957	51970
1	1983	1996	2009	2022	2035	2048	2061	2075	2088	2101
2	2114	2127	2140	2153	2166	2179	2192	2205	2218	2231
3	2244	2257	2270	2284	2297	2310	2323	2336	2349	2362
4	2375	2388	2401	2414	2427	2440	2453	2466	2479	2492
5	2504	2517	2530	2543	2556	2569	2582	2595	2608	2621
6	2634	2647	2660	2673	2686	2699	2711	2724	2737	2750
7	2763	2776	2789	2802	2815	2827	2840	2853	2866	2879
8	2892	2905	2917	2930	2943	2956	2969	2982	2994	3007
9	3020	3033	3046	3058	3071	3084	3097	3110	3122	3135
340	53148	53161	53173	53186	53199	53212	53224	53237	53250	53263
1	3275	3288	3301	3314	3326	3339	3352	3364	3377	3390
2	3403	3415	3428	3441	3453	3466	3479	3491	3504	3517
3	3529	3542	3555	3567	3580	3593	3605	3618	3631	3643
4	3656	3668	3681	3694	3706	3719	3732	3744	3757	3769
5	3782	3794	3807	3820	3832	3845	3857	3870	3882	3895
6	3908	3920	3933	3945	3958	3970	3983	3995	4008	4020
7	4033	4045	4058	4070	4083	4095	4108	4120	4133	4145
8	4158	4170	4183	4195	4208	4220	4233	4245	4258	4270
9	4283	4295	4307	4320	4332	4345	4357	4370	4382	4394
350	54407	54419	54432	54444	54456	54469	54481	54494	54506	54518

TABLE 1. LOGARITHMS OF NUMBERS

N	0	1	2	3	4	5	6	7	8	9
350	54407	54419	54432	54444	54456	54469	54481	54494	54506	54518
1	4531	4543	4555	4568	4580	4593	4605	4617	4630	4642
2	4654	4667	4679	4691	4704	4716	4728	4741	4753	4765
3	4777	4790	4802	4814	4827	4839	4851	4864	4876	4888
4	4900	4913	4925	4937	4949	4962	4974	4986	4998	5011
5	5023	5035	5047	5060	5072	5084	5096	5108	5121	5133
6	5145	5157	5169	5182	5194	5206	5218	5230	5242	5255
7	5267	5279	5291	5303	5315	5328	5340	5352	5364	5376
8	5388	5400	5413	5425	5437	5449	5461	5473	5485	5497
9	5509	5522	5534	5546	5558	5570	5582	5594	5606	5618
360	55630	55642	55654	55666	55678	55691	55703	55715	55727	55739
1	5751	5763	5775	5787	5799	5811	5823	5835	5847	5859
2	5871	5883	5895	5907	5919	5931	5943	5955	5967	5979
3	5991	6003	6015	6027	6038	6050	6062	6074	6086	6098
4	6110	6122	6134	6146	6158	6170	6182	6194	6205	6217
5	6229	6241	6253	6265	6277	6289	6301	6312	6324	6336
6	6348	6360	6372	6384	6396	6407	6419	6431	6443	6455
7	6467	6478	6490	6502	6514	6526	6538	6549	6561	6573
8	6585	6597	6608	6620	6632	6644	6656	6667	6679	6691
9	6703	6714	6726	6738	6750	6761	6773	6785	6797	6808
370	56820	56832	56844	56855	56867	56879	56891	56902	56914	56926
1	6937	6949	6961	6972	6984	6996	7008	7019	7031	7043
2	7054	7066	7078	7089	7101	7113	7124	7136	7148	7159
3	7171	7183	7194	7206	7217	7229	7241	7252	7264	7276
4	7287	7299	7310	7322	7334	7345	7357	7368	7380	7392
5	7403	7415	7426	7438	7449	7461	7473	7484	7496	7507
6	7519	7530	7542	7553	7565	7576	7588	7600	7611	7623
7	7634	7646	7657	7669	7680	7692	7703	7715	7726	7738
8	7749	7761	7772	7784	7795	7807	7818	7830	7841	7852
9	7864	7875	7887	7898	7910	7921	7933	7944	7955	7967
380	57978	57990	58001	58013	58024	58035	58047	58058	58070	58081
1	8092	8104	8115	8127	8138	8149	8161	8172	8184	8195
2	8206	8218	8229	8240	8252	8263	8274	8286	8297	8309
3	8320	8331	8343	8354	8365	8377	8388	8399	8410	8422
4	8433	8444	8456	8467	8478	8490	8501	8512	8524	8535
5	8546	8557	8569	8580	8591	8602	8614	8625	8636	8647
6	8659	8670	8681	8692	8704	8715	8726	8737	8749	8760
7	8771	8782	8794	8805	8816	8827	8838	8850	8861	8872
8	8883	8894	8906	8917	8928	8939	8950	8961	8973	8984
9	8995	9006	9017	9028	9040	9051	9062	9073	9084	9095
390	59106	59118	59129	59140	59151	59162	59173	59184	59195	59207
1	9218	9229	9240	9251	9262	9273	9284	9295	9306	9318
2	9329	9340	9351	9362	9373	9384	9395	9406	9417	9428
3	9439	9450	9461	9472	9483	9494	9506	9517	9528	9539
4	9550	9561	9572	9583	9594	9605	9616	9627	9638	9649
5	9660	9671	9682	9693	9704	9715	9726	9737	9748	9759
6	9770	9780	9791	9802	9813	9824	9835	9846	9857	9868
7	9879	9890	9901	9912	9923	9934	9945	9956	9966	9977
8	9988	9999	60010	60021	60032	60043	60054	60065	60076	60086
9	60097	60108	0119	0130	0141	0152	0163	0173	0184	0195
400	60206	60217	60228	60239	60249	60260	60271	60282	60293	60304

TABLE 1. LOGARITHMS OF NUMBERS

N	0	1	2	3	4	5	6	7	8	9
400	60206	60217	60228	60239	60249	60260	60271	60282	60293	60304
1	0314	0325	0336	0347	0358	0369	0379	0390	0401	0412
2	0423	0433	0444	0455	0466	0477	0487	0498	0509	0520
3	0531	0541	0552	0563	0574	0584	0595	0606	0617	0627
4	0638	0649	0660	0670	0681	0692	0703	0713	0724	0735
5	0746	0756	0767	0778	0788	0799	0810	0821	0831	0842
6	0853	0863	0874	0885	0895	0906	0917	0927	0938	0949
7	0959	0970	0981	0991	1002	1013	1023	1034	1045	1055
8	1066	1077	1087	1098	1109	1119	1130	1140	1151	1162
9	1172	1183	1194	1204	1215	1225	1236	1247	1257	1268
410	61278	61289	61300	61310	61321	61331	61342	61352	61363	61374
1	1384	1395	1405	1416	1426	1437	1448	1458	1469	1479
2	1490	1500	1511	1521	1532	1542	1553	1563	1574	1584
3	1595	1606	1616	1627	1637	1648	1658	1669	1679	1690
4	1700	1711	1721	1731	1742	1752	1763	1773	1784	1794
5	1805	1815	1826	1836	1847	1857	1868	1878	1888	1899
6	1909	1920	1930	1941	1951	1962	1972	1982	1993	2003
7	2014	2024	2034	2045	2055	2066	2076	2086	2097	2107
8	2118	2128	2138	2149	2159	2170	2180	2190	2201	2211
9	2221	2232	2242	2252	2263	2273	2284	2294	2304	2315
420	62325	62335	62346	62356	62366	62377	62387	62397	62408	62418
1	2428	2439	2449	2459	2469	2480	2490	2500	2511	2521
2	2531	2542	2552	2562	2572	2583	2593	2603	2613	2624
3	2634	2644	2655	2665	2675	2685	2696	2706	2716	2726
4	2737	2747	2757	2767	2778	2788	2798	2808	2818	2829
5	2839	2849	2859	2870	2880	2890	2900	2910	2921	2931
6	2941	2951	2961	2972	2982	2992	3002	3012	3022	3033
7	3043	3053	3063	3073	3083	3094	3104	3114	3124	3134
8	3144	3155	3165	3175	3185	3195	3205	3215	3225	3236
9	3246	3256	3266	3276	3286	3296	3306	3317	3327	3337
430	63347	63357	63367	63377	63387	63397	63407	63417	63428	63438
1	3448	3458	3468	3478	3488	3498	3508	3518	3528	3538
2	3548	3558	3568	3579	3589	3599	3609	3619	3629	3639
3	3649	3659	3669	3679	3689	3699	3709	3719	3729	3739
4	3749	3759	3769	3779	3789	3799	3809	3819	3829	3839
5	3849	3859	3869	3879	3889	3899	3909	3919	3929	3939
6	3949	3959	3969	3979	3988	3998	4008	4018	4028	4038
7	4048	4058	4068	4078	4088	4098	4108	4118	4128	4137
8	4147	4157	4167	4177	4187	4197	4207	4217	4227	4237
9	4246	4256	4266	4276	4286	4296	4306	4316	4326	4335
440	64345	64355	64365	64375	64385	64395	64404	64414	64424	64434
1	4444	4454	4464	4473	4483	4493	4503	4513	4523	4532
2	4542	4552	4562	4572	4582	4591	4601	4611	4621	4631
3	4640	4650	4660	4670	4680	4689	4699	4709	4719	4729
4	4738	4748	4758	4768	4777	4787	4797	4807	4816	4826
5	4836	4846	4856	4865	4875	4885	4895	4904	4914	4924
6	4933	4943	4953	4963	4972	4982	4992	5002	5011	5021
7	5031	5040	5050	5060	5070	5079	5089	5099	5108	5118
8	5128	5137	5147	5157	5167	5176	5186	5196	5205	5215
9	5225	5234	5244	5254	5263	5273	5283	5292	5302	5312
450	65321	65331	65341	65350	65360	65369	65379	65389	65398	65408

TABLE 1. LOGARITHMS OF NUMBERS

N	0	1	2	3	4	5	6	7	8	9
450	65321	65331	65341	65350	65360	65369	65379	65389	65398	65408
1	5418	5427	5437	5447	5456	5466	5475	5485	5495	5504
2	5514	5523	5533	5543	5552	5562	5571	5581	5591	5600
3	5610	5619	5629	5639	5648	5658	5667	5677	5686	5696
4	5706	5715	5725	5734	5744	5753	5763	5772	5782	5792
5	5801	5811	5820	5830	5839	5849	5858	5868	5877	5887
6	5896	5906	5916	5925	5935	5944	5954	5963	5973	5982
7	5992	6001	6011	6020	6030	6039	6049	6058	6068	6077
8	6087	6096	6106	6115	6124	6134	6143	6153	6162	6172
9	6181	6191	6200	6210	6219	6229	6238	6247	6257	6266
460	66276	66285	66295	66304	66314	66323	66333	66342	66351	66361
1	6370	6380	6389	6398	6408	6417	6427	6436	6445	6455
2	6464	6474	6483	6492	6502	6511	6521	6530	6539	6549
3	6558	6567	6577	6586	6596	6605	6614	6624	6633	6642
4	6652	6661	6671	6680	6689	6699	6708	6717	6727	6736
5	6745	6755	6764	6773	6783	6792	6801	6811	6820	6829
6	6839	6848	6857	6867	6876	6885	6894	6904	6913	6922
7	6932	6941	6950	6960	6969	6978	6987	6997	7006	7015
8	7025	7034	7043	7052	7062	7071	7080	7089	7099	7108
9	7117	7127	7136	7145	7154	7164	7173	7182	7191	7201
470	67210	67219	67228	67237	67247	67256	67265	67274	67284	67293
1	7302	7311	7321	7330	7339	7348	7357	7367	7376	7385
2	7394	7403	7413	7422	7431	7440	7449	7459	7468	7477
3	7486	7495	7504	7514	7523	7532	7541	7550	7560	7569
4	7578	7587	7596	7605	7614	7624	7633	7642	7651	7660
5	7669	7679	7688	7697	7706	7715	7724	7733	7742	7752
6	7761	7770	7779	7788	7797	7806	7815	7825	7834	7843
7	7852	7861	7870	7879	7888	7897	7906	7916	7925	7934
8	7943	7952	7961	7970	7979	7988	7997	8006	8015	8024
9	8034	8043	8052	8061	8070	8079	8088	8097	8106	8115
480	68124	68133	68142	68151	68160	68169	68178	68187	68196	68205
1	8215	8224	8233	8242	8251	8260	8269	8278	8287	8296
2	8305	8314	8323	8332	8341	8350	8359	8368	8377	8386
3	8395	8404	8413	8422	8431	8440	8449	8458	8467	8476
4	8485	8494	8502	8511	8520	8529	8538	8547	8556	8565
5	8574	8583	8592	8601	8610	8619	8628	8637	8646	8655
6	8664	8673	8681	8690	8699	8708	8717	8726	8735	8744
7	8753	8762	8771	8780	8789	8797	8806	8815	8824	8833
8	8842	8851	8860	8869	8878	8886	8895	8904	8913	8922
9	8931	8940	8949	8958	8966	8975	8984	8993	9002	9011
490	69020	69028	69037	69046	69055	69064	69073	69082	69090	69099
1	9108	9117	9126	9135	9144	9152	9161	9170	9179	9188
2	9197	9205	9214	9223	9232	9241	9249	9258	9267	9276
3	9285	9294	9302	9311	9320	9329	9338	9346	9355	9364
4	9373	9381	9390	9399	9408	9417	9425	9434	9443	9452
5	9461	9469	9478	9487	9496	9504	9513	9522	9531	9539
6	9548	9557	9566	9574	9583	9592	9601	9609	9618	9627
7	9636	9644	9653	9662	9671	9679	9688	9697	9705	9714
8	9723	9732	9740	9749	9758	9767	9775	9784	9793	9801
9	9810	9819	9827	9836	9845	9854	9862	9871	9880	9888
500	69897	69906	69914	69923	69932	69940	69949	69958	69966	69975

TABLE 1. LOGARITHMS OF NUMBERS

N	0	1	2	3	4	5	6	7	8	9
500	69897	69906	69914	69923	69932	69940	69949	69958	69966	69975
1	9984	9992	70001	70010	70018	70027	70036	70044	70053	70062
2	70070	70079	0088	0096	0105	0114	0122	0131	0140	0148
3	0157	0165	0174	0183	0191	0200	0209	0217	0226	0234
4	0243	0252	0260	0269	0278	0286	0295	0303	0312	0321
5	0329	0338	0346	0355	0364	0372	0381	0389	0398	0406
6	0415	0424	0432	0441	0449	0458	0467	0475	0484	0492
7	0501	0509	0518	0526	0535	0544	0552	0561	0569	0578
8	0586	0595	0603	0612	0621	0629	0638	0646	0655	0663
9	0672	0680	0689	0697	0706	0714	0723	0731	0740	0749
510	70757	70766	70774	70783	70791	70800	70808	70817	70825	70834
1	0842	0851	0859	0868	0876	0885	0893	0902	0910	0919
2	0927	0935	0944	0952	0961	0969	0978	0986	0995	1003
3	1012	1020	1029	1037	1046	1054	1063	1071	1079	1088
4	1096	1105	1113	1122	1130	1139	1147	1155	1164	1172
5	1181	1189	1198	1206	1214	1223	1231	1240	1248	1257
6	1265	1273	1282	1290	1299	1307	1315	1324	1332	1341
7	1349	1357	1366	1374	1383	1391	1399	1408	1416	1425
8	1433	1441	1450	1458	1466	1475	1483	1492	1500	1508
9	1517	1525	1533	1542	1550	1559	1567	1575	1584	1592
520	71600	71609	71617	71625	71634	71642	71650	71659	71667	71675
1	1684	1692	1700	1709	1717	1725	1734	1742	1750	1759
2	1767	1775	1784	1792	1800	1809	1817	1825	1834	1842
3]	1850	1858	1867	1875	1883	1892	1900	1908	1917	1925
4	1933	1941	1950	1958	1966	1975	1983	1991	1999	2008
5	2016	2024	2032	2041	2049	2057	2066	2074	2082	2090
6	2099	2107	2115	2123	2132	2140	2148	2156	2165	2173
7	2181	2189	2198	2206	2214	2222	2230	2239	2247	2255
8	2263	2272	2280	2288	2296	2304	2313	2321	2329	2337
9	2346	2354	2362	2370	2378	2387	2395	2403	2411	2419
530	72428	72436	72444	72452	72460	72469	72477	72485	72493	72501
1	2509	2518	2526	2534	2542	2550	2558	2567	2575	2583
2	2591	2599	2607	2616	2624	2632	2640	2648	2656	2665
3	2673	2681	2689	2697	2705	2713	2722	2730	2738	2746
4	2754	2762	2770	2779	2787	2795	2803	2811	2819	2827
5	2835	2843	2852	2860	2868	2876	2884	2892	2900	2908
6	2916	2925	2933	2941	2949	2957	2965	2973	2981	2989
7	2997	3006	3014	3022	3030	3038	3046	3054	3062	3070
8	3078	3086	3094	3102	3111	3119	3127	3135	3143	3151
9	3159	3167	3175	3183	3191	3199	3207	3215	3223	3231
540	73239	73247	73255	73263	73272	73280	73288	73296	73304	73312
1	3320	3328	3336	3344	3352	3360	3368	3376	3384	3392
2	3400	3408	3416	3424	3432	3440	3448	3456	3464	3472
3	3480	3488	3496	3504	3512	3520	3528	3536	3544	3552
4	3560	3568	3576	3584	3592	3600	3608	3616	3624	3632
5	3640	3648	3656	3664	3672	3679	3687	3695	3703	3711
6	3719	3727	3735	3743	3751	3759	3767	3775	3783	3791
7	3799	3807	3815	3823	3830	3838	3846	3854	3862	3870
8	3878	3886	3894	3902	3910	3918	3926	3933	3941	3949
9	3957	3965	3973	3981	3989	3997	4005	4013	4020	4028
550	74036	74044	74052	74060	74068	74076	74084	74092	74099	74107

TABLE 1. LOGARITHMS OF NUMBERS

N	0	1	2	3	4	5	6	7	8	9
550	74036	74044	74052	74060	74068	74076	74084	74092	74099	74107
1	4115	4123	4131	4139	4147	4155	4162	4170	4178	4186
2	4194	4202	4210	4218	4225	4233	4241	4249	4257	4265
3	4273	4280	4288	4296	4304	4312	4320	4327	4335	4343
4	4351	4359	4367	4374	4382	4390	4398	4406	4414	4421
5	4429	4437	4445	4453	4461	4468	4476	4484	4492	4500
6	4507	4515	4523	4531	4539	4547	4554	4562	4570	4578
7	4586	4593	4601	4609	4617	4624	4632	4640	4648	4656
8	4663	4671	4679	4687	4695	4702	4710	4718	4726	4733
9	4741	4749	4757	4764	4772	4780	4788	4796	4803	4811
560	74819	74827	74834	74842	74850	74858	74865	74873	74881	74889
1	4896	4904	4912	4920	4927	4935	4943	4950	4958	4966
2	4974	4981	4989	4997	5005	5012	5020	5028	5035	5043
3	5051	5059	5066	5074	5082	5089	5097	5105	5113	5120
4	5128	5136	5143	5151	5159	5166	5174	5182	5189	5197
5	5205	5213	5220	5228	5236	5243	5251	5259	5266	5274
6	5282	5289	5297	5305	5312	5320	5328	5335	5343	5351
7	5358	5366	5374	5381	5389	5397	5404	5412	5420	5427
8	5435	5442	5450	5458	5465	5473	5481	5488	5496	5504
9	5511	5519	5526	5534	5542	5549	5557	5565	5572	5580
570	75587	75595	75603	75610	75618	75626	75633	75641	75648	75656
1	5664	5671	5679	5686	5694	5702	5709	5717	5724	5732
2	5740	5747	5755	5762	5770	5778	5785	5793	5800	5808
3	5815	5823	5831	5838	5846	5853	5861	5868	5876	5884
4	5891	5899	5906	5914	5921	5929	5937	5944	5952	5959
5	5967	5974	5982	5989	5997	6005	6012	6020	6027	6035
6	6042	6050	6057	6065	6072	6080	6087	6095	6103	6110
7	6118	6125	6133	6140	6148	6155	6163	6170	6178	6185
8	6193	6200	6208	6215	6223	6230	6238	6245	6253	6260
9	6268	6275	6283	6290	6298	6305	6313	6320	6328	6335
580	76343	76350	76358	76365	76373	76380	76388	76395	76403	76410
1	6418	6425	6433	6440	6448	6455	6462	6470	6477	6485
2	6492	6500	6507	6515	6522	6530	6537	6545	6552	6559
3	6567	6574	6582	6589	6597	6604	6612	6619	6626	6634
4	6641	6649	6656	6664	6671	6678	6686	6693	6701	6708
5	6716	6723	6730	6738	6745	6753	6760	6768	6775	6782
6	6790	6797	6805	6812	6819	6827	6834	6842	6849	6856
7	6864	6871	6879	6886	6893	6901	6908	6916	6923	6930
8	6938	6945	6953	6960	6967	6975	6982	6989	6997	7004
9	7012	7019	7026	7034	7041	7048	7056	7063	7070	7078
590	77085	77093	77100	77107	77115	77122	77129	77137	77144	77151
1	7159	7166	7173	7181	7188	7195	7203	7210	7217	7225
2	7232	7240	7247	7254	7262	7269	7276	7283	7291	7298
3	7305	7313	7320	7327	7335	7342	7349	7357	7364	7371
4	7379	7386	7393	7401	7408	7415	7422	7430	7437	7444
5	7452	7459	7466	7474	7481	7488	7495	7503	7510	7517
6	7525	7532	7539	7546	7554	7561	7568	7576	7583	7590
7	7597	7605	7612	7619	7627	7634	7641	7648	7656	7663
8	7670	7677	7685	7692	7699	7706	7714	7721	7728	7735
9	7743	7750	7757	7764	7772	7779	7786	7793	7801	7808
600	77815	77822	77830	77837	77844	77851	77859	77866	77873	77880

TABLE 1. LOGARITHMS OF NUMBERS

N	0	1	2	3	4	5	6	7	8	9
600	77815	77822	77830	77837	77844	77851	77859	77866	77873	77880
1	7887	7895	7902	7909	7916	7924	7931	7938	7945	7952
2	7960	7967	7974	7981	7988	7996	8003	8010	8017	8025
3	8032	8039	8046	8053	8061	8068	8075	8082	8089	8097
4	8104	8111	8118	8125	8132	8140	8147	8154	8161	8168
5	8176	8183	8190	8197	8204	8211	8219	8226	8233	8240
6	8247	8254	8262	8269	8276	8283	8290	8297	8305	8312
7	8319	8326	8333	8340	8347	8355	8362	8369	8376	8383
8	8390	8398	8405	8412	8419	8426	8433	8440	8447	8455
9	8462	8469	8476	8483	8490	8497	8504	8512	8519	8526
610	78533	78540	78547	78554	78561	78569	78576	78583	78590	78597
1	8604	8611	8618	8625	8633	8640	8647	8654	8661	8668
2	8675	8682	8689	8696	8704	8711	8718	8725	8732	8739
3	8746	8753	8760	8767	8774	8781	8789	8796	8803	8810
4	8817	8824	8831	8838	8845	8852	8859	8866	8873	8880
5	8888	8895	8902	8909	8916	8923	8930	8937	8944	8951
6	8958	8965	8972	8979	8986	8993	9000	9007	9014	9021
7	9029	9036	9043	9050	9057	9064	9071	9078	9085	9092
8	9099	9106	9113	9120	9127	9134	9141	9148	9155	9162
9	9169	9176	9183	9190	9197	9204	9211	9218	9225	9232
620	79239	79246	79253	79260	79267	79274	79281	79288	79295	79302
1	9309	9316	9323	9330	9337	9344	9351	9358	9365	9372
2	9379	9386	9393	9400	9407	9414	9421	9428	9435	9442
3	9449	9456	9463	9470	9477	9484	9491	9498	9505	9511
4	9518	9525	9532	9539	9546	9553	9560	9567	9574	9581
5	9588	9595	9602	9609	9616	9623	9630	9637	9644	9650
6	9657	9664	9671	9678	9685	9692	9699	9706	9713	9720
7	9727	9734	9741	9748	9754	9761	9768	9775	9782	9789
8	9796	9803	9810	9817	9824	9831	9837	9844	9851	9858
9	9865	9872	9879	9886	9893	9900	9906	9913	9920	9927
630	79934	79941	79948	79955	79962	79969	79975	79982	79989	79996
1	80003	80010	80017	80024	80030	80037	80044	80051	80058	80065
2	0072	0079	0085	0092	0099	0106	0113	0120	0127	0134
3	0140	0147	0154	0161	0168	0175	0182	0188	0195	0202
4	0209	0216	0223	0229	0236	0243	0250	0257	0264	0271
5	0277	0284	0291	0298	0305	0312	0318	0325	0332	0339
6	0346	0353	0359	0366	0373	0380	0387	0393	0400	0407
7	0414	0421	0428	0434	0441	0448	0455	0462	0468	0475
8	0482	0489	0496	0502	0509	0516	0523	0530	0536	0543
9	0550	0557	0564	0570	0577	0584	0591	0598	0604	0611
640	80618	80625	80632	80638	80645	80652	80659	80665	80672	80679
1	0686	0693	0699	0706	0713	0720	0726	0733	0740	0747
2	0754	0760	0767	0774	0781	0787	0794	0801	0808	0814
3	0821	0828	0835	0841	0848	0855	0862	0868	0875	0882
4	0889	0895	0902	0909	0916	0922	0929	0936	0943	0949
5	0956	0963	0969	0976	0983	0990	0996	1003	1010	1017
6	1023	1030	1037	1043	1050	1057	1064	1070	1077	1084
7	1090	1097	1104	1111	1117	1124	1131	1137	1144	1151
8	1158	1164	1171	1178	1184	1191	1198	1204	1211	1218
9	1224	1231	1238	1245	1251	1258	1265	1271	1278	1285
650	81291	81298	81305	81311	81318	81325	81331	81338	81345	81351

TABLE 1. LOGARITHMS OF NUMBERS

N	0	1	2	3	4	5	6	7	8	9
650	81291	81298	81305	81311	81318	81325	81331	81338	81345	81351
1	1358	1365	1371	1378	1385	1391	1398	1405	1411	1418
2	1425	1431	1438	1445	1451	1458	1465	1471	1478	1485
3	1491	1498	1505	1511	1518	1525	1531	1538	1544	1551
4	1558	1564	1571	1578	1584	1591	1598	1604	1611	1617
5	1624	1631	1637	1644	1651	1657	1664	1671	1677	1684
6	1690	1697	1704	1710	1717	1723	1730	1737	1743	1750
7	1757	1763	1770	1776	1783	1790	1796	1803	1809	1816
8	1823	1829	1836	1842	1849	1856	1862	1869	1875	1882
9	1889	1895	1902	1908	1915	1921	1928	1935	1941	1948
660	81954	81961	81968	81974	81981	81987	81994	82000	82007	82014
1	2020	2027	2033	2040	2046	2053	2060	2066	2073	2079
2	2086	2092	2099	2105	2112	2119	2125	2132	2138	2145
3	2151	2158	2164	2171	2178	2184	2191	2197	2204	2210
4	2217	2223	2230	2236	2243	2249	2256	2263	2269	2276
5	2282	2289	2295	2302	2308	2315	2321	2328	2334	2341
6	2347	2354	2360	2367	2373	2380	2387	2393	2400	2406
7	2413	2419	2426	2432	2439	2445	2452	2458	2465	2471
8	2478	2484	2491	2497	2504	2510	2517	2523	2530	2536
9	2543	2549	2556	2562	2569	2575	2582	2588	2595	2601
670	82607	82614	82620	82627	82633	82640	82646	82653	82659	82666
1	2672	2679	2685	2692	2698	2705	2711	2718	2724	2730
2	2737	2743	2750	2756	2763	2769	2776	2782	2789	2795
3	2802	2808	2814	2821	2827	2834	2840	2847	2853	2860
4	2866	2872	2879	2885	2892	2898	2905	2911	2918	2924
5	2930	2937	2943	2950	2956	2963	2969	2975	2982	2988
6	2995	3001	3008	3014	3020	3027	3033	3040	3046	3052
7	3059	3065	3072	3078	3085	3091	3097	3104	3110	3117
8	3123	3129	3136	3142	3149	3155	3161	3168	3174	3181
9	3187	3193	3200	3206	3213	3219	3225	3232	3238	3245
680	83251	83257	83264	83270	83276	83283	83289	83296	83302	83308
1	3315	3321	3327	3334	3340	3347	3353	3359	3366	3372
2	3378	3385	3391	3398	3404	3410	3417	3423	3429	3436
3	3442	3448	3455	3461	3467	3474	3480	3487	3493	3499
4	3506	3512	3518	3525	3531	3537	3544	3550	3556	3563
5	3569	3575	3582	3588	3594	3601	3607	3613	3620	3626
6	3632	3639	3645	3651	3658	3664	3670	3677	3683	3689
7	3696	3702	3708	3715	3721	3727	3734	3740	3746	3753
8	3759	3765	3771	3778	3784	3790	3797	3803	3809	3816
9	3822	3828	3835	3841	3847	3853	3860	3866	3872	3879
690	83885	83891	83897	83904	83910	83916	83923	83929	83935	83942
1	3948	3954	3960	3967	3973	3979	3985	3992	3998	4004
2	4011	4017	4023	4029	4036	4042	4048	4055	4061	4067
3	4073	4080	4086	4092	4098	4105	4111	4117	4123	4130
4	4136	4142	4148	4155	4161	4167	4173	4180	4186	4192
5	4198	4205	4211	4217	4223	4230	4236	4242	4248	4255
6	4261	4267	4273	4280	4286	4292	4298	4305	4311	4317
7	4323	4330	4336	4342	4348	4354	4361	4367	4373	4379
8	4386	4392	4398	4404	4410	4417	4423	4429	4435	4442
9	4448	4454	4460	4466	4473	4479	4485	4491	4497	4504
700	84510	84516	84522	84528	84535	84541	84547	84553	84559	84566

TABLE 1. LOGARITHMS OF NUMBERS

N	0	1	2	3	4	5	6	7	8	9
700	84510	84516	84522	84528	84535	84541	84547	84553	84559	84566
1	4572	4578	4584	4590	4597	4603	4609	4615	4621	4628
2	4634	4640	4646	4652	4658	4665	4671	4677	4683	4689
3	4696	4702	4708	4714	4720	4726	4733	4739	4745	4751
4	4757	4763	4770	4776	4782	4788	4794	4800	4807	4813
5	4819	4825	4831	4837	4844	4850	4856	4862	4868	4874
6	4880	4887	4893	4899	4905	4911	4917	4924	4930	4936
7	4942	4948	4954	4960	4967	4973	4979	4985	4991	4997
8	5003	5009	5016	5022	5028	5034	5040	5046	5052	5058
9	5065	5071	5077	5083	5089	5095	5101	5107	5114	5120
710	85126	85132	85138	85144	85150	85156	85163	85169	85175	85181
1	5187	5193	5199	5205	5211	5217	5224	5230	5236	5242
2	5248	5254	5260	5266	5272	5278	5285	5291	5297	5303
3	5309	5315	5321	5327	5333	5339	5345	5352	5358	5364
4	5370	5376	5382	5388	5394	5400	5406	5412	5418	5425
5	5431	5437	5443	5449	5455	5461	5467	5473	5479	5485
6	5491	5497	5503	5509	5516	5522	5528	5534	5540	5546
7	5552	5558	5564	5570	5576	5582	5588	5594	5600	5606
8	5612	5618	5625	5631	5637	5643	5649	5655	5661	5667
9	5673	5679	5685	5691	5697	5703	5709	5715	5721	5727
720	85733	85739	85745	85751	85757	85763	85769	85775	85781	85788
1	5794	5800	5806	5812	5818	5824	5830	5836	5842	5848
2	5854	5860	5866	5872	5878	5884	5890	5896	5902	5908
3	5914	5920	5926	5932	5938	5944	5950	5956	5962	5968
4	5974	5980	5986	5992	5998	6004	6010	6016	6022	6028
5	6034	6040	6046	6052	6058	6064	6070	6076	6082	6088
6	6094	6100	6106	6112	6118	6124	6130	6136	6141	6147
7	6153	6159	6165	6171	6177	6183	6189	6195	6201	6207
8	6213	6219	6225	6231	6237	6243	6249	6255	6261	6267
9	6273	6279	6285	6291	6297	6303	6308	6314	6320	6326
730	86332	86338	86344	86350	86356	86362	86368	86374	86380	86386
1	6392	6398	6404	6410	6415	6421	6427	6433	6439	6445
2	6451	6457	6463	6469	6475	6481	6487	6493	6499	6504
3	6510	6516	6522	6528	6534	6540	6546	6552	6558	6564
4	6570	6576	6581	6587	6593	6599	6605	6611	6617	6623
5	6629	6635	6641	6646	6652	6658	6664	6670	6676	6682
6	6688	6694	6700	6705	6711	6717	6723	6729	6735	6741
7	6747	6753	6759	6764	6770	6776	6782	6788	6794	6800
8	6806	6812	6817	6823	6829	6835	6841	6847	6853	6859
9	6864	6870	6876	6882	6888	6894	6900	6906	6911	6917
740	86923	86929	86935	86941	86947	86953	86958	86964	86970	86976
1	6982	6988	6994	6999	7005	7011	7017	7023	7029	7035
2	7040	7046	7052	7058	7064	7070	7075	7081	7087	7093
3	7099	7105	7111	7116	7122	7128	7134	7140	7146	7151
4	7157	7163	7169	7175	7181	7186	7192	7198	7204	7210
5	7216	7221	7227	7233	7239	7245	7251	7256	7262	7268
6	7274	7280	7286	7291	7297	7303	7309	7315	7320	7326
7	7332	7338	7344	7349	7355	7361	7367	7373	7379	7384
8	7390	7396	7402	7408	7413	7419	7425	7431	7437	7442
9	7448	7454	7460	7466	7471	7477	7483	7489	7495	7500
750	87506	87512	87518	87523	87529	87535	87541	87547	87552	87558

TABLE 1. LOGARITHMS OF NUMBERS

N	0	1	2	3	4	5	6	7	8	9
750	87506	87512	87518	87523	87529	87535	87541	87547	87552	87558
1	7564	7570	7576	7581	7587	7593	7599	7604	7610	7616
2	7622	7628	7633	7639	7645	7651	7656	7662	7668	7674
3	7679	7685	7691	7697	7703	7708	7714	7720	7726	7731
4	7737	7743	7749	7754	7760	7766	7772	7777	7783	7789
5	7795	7800	7806	7812	7818	7823	7829	7835	7841	7846
6	7852	7858	7864	7869	7875	7881	7887	7892	7898	7904
7	7910	7915	7921	7927	7933	7938	7944	7950	7955	7961
8	7967	7973	7978	7984	7990	7996	8001	8007	8013	8018
9	8024	8030	8036	8041	8047	8053	8058	8064	8070	8076
760	88081	88087	88093	88098	88104	88110	88116	88121	88127	88133
1	8138	8144	8150	8156	8161	8167	8173	8178	8184	8190
2	8195	8201	8207	8213	8218	8224	8230	8235	8241	8247
3	8252	8258	8264	8270	8275	8281	8287	8292	8298	8304
4	8309	8315	8321	8326	8332	8338	8343	8349	8355	8360
5	8366	8372	8377	8383	8389	8395	8400	8406	8412	8417
6	8423	8429	8434	8440	8446	8451	8457	8463	8468	8474
7	8480	8485	8491	8497	8502	8508	8513	8519	8525	8530
8	8536	8542	8547	8553	8559	8564	8570	8576	8581	8587
9	8593	8598	8604	8610	8615	8621	8627	8632	8638	8643
770	88649	88655	88660	88666	88672	88677	88683	88689	88694	88700
1	8705	8711	8717	8722	8728	8734	8739	8745	8750	8756
2	8762	8767	8773	8779	8784	8790	8795	8801	8807	8812
3	8818	8824	8829	8835	8840	8846	8852	8857	8863	8868
4	8874	8880	8885	8891	8897	8902	8908	8913	8919	8925
5	8930	8936	8941	8947	8953	8958	8964	8969	8975	8981
6	8986	8992	8997	9003	9009	9014	9020	9025	9031	9037
7	9042	9048	9053	9059	9064	9070	9076	9081	9087	9092
8	9098	9104	9109	9115	9120	9126	9131	9137	9143	9148
9	9154	9159	9165	9170	9176	9182	9187	9193	9198	9204
780	89209	89215	89221	89226	89232	89237	89243	89248	89254	89260
1	9265	9271	9276	9282	9287	9293	9298	9304	9310	9315
2	9321	9326	9332	9337	9343	9348	9354	9360	9365	9371
3	9376	9382	9387	9393	9398	9404	9409	9415	9421	9426
4	9432	9437	9443	9448	9454	9459	9465	9470	9476	9481
5	9487	9492	9498	9504	9509	9515	9520	9526	9531	9537
6	9542	9548	9553	9559	9564	9570	9575	9581	9586	9592
7	9597	9603	9609	9614	9620	9625	9631	9636	9642	9647
8	9653	9658	9664	9669	9675	9680	9686	9691	9697	9702
9	9708	9713	9719	9724	9730	9735	9741	9746	9752	9757
790	89763	89768	89774	89779	89785	89790	89796	89801	89807	89812
1	9818	9823	9829	9834	9840	9845	9851	9856	9862	9867
2	9873	9878	9883	9889	9894	9900	9905	9911	9916	9922
3	9927	9933	9938	9944	9949	9955	9960	9966	9971	9977
4	9982	9988	9993	9998	90004	90009	90015	90020	90026	90031
5	90037	90042	90048	90053	0059	0064	0069	0075	0080	0086
6	0091	0097	0102	0108	0113	0119	0124	0129	0135	0140
7	0146	0151	0157	0162	0168	0173	0179	0184	0189	0195
8	0200	0206	0211	0217	0222	0227	0233	0238	0244	0249
9	0255	0260	0266	0271	0276	0282	0287	0293	0298	0304
800	90309	90314	90320	90325	90331	90336	90342	90347	90352	90358

TABLE 1. LOGARITHMS OF NUMBERS

N	0	1	2	3	4	5	6	7	8	9
800	90309	90314	90320	90325	90331	90336	90342	90347	90352	90358
1	0363	0369	0374	0380	0385	0390	0396	0401	0407	0412
2	0417	0423	0428	0434	0439	0445	0450	0455	0461	0466
3	0472	0477	0482	0488	0493	0499	0504	0509	0515	0520
4	0526	0531	0536	0542	0547	0553	0558	0563	0569	0574
5	0580	0585	0590	0596	0601	0607	0612	0617	0623	0628
6	0634	0639	0644	0650	0655	0660	0666	0671	0677	0682
7	0687	0693	0698	0703	0709	0714	0720	0725	0730	0736
8	0741	0747	0752	0757	0763	0768	0773	0779	0784	0789
9	0795	0800	0806	0811	0816	0822	0827	0832	0838	0843
810	90849	90854	90859	90865	90870	90875	90881	90886	90891	90897
1	0902	0907	0913	0918	0924	0929	0934	0940	0945	0950
2	0956	0961	0966	0972	0977	0982	0988	0993	0998	1004
3	1009	1014	1020	1025	1030	1036	1041	1046	1052	1057
4	1062	1068	1073	1078	1084	1089	1094	1100	1105	1110
5	1116	1121	1126	1132	1137	1142	1148	1153	1158	1164
6	1169	1174	1180	1185	1190	1196	1201	1206	1212	1217
7	1222	1228	1233	1238	1243	1249	1254	1259	1265	1270
8	1275	1281	1286	1291	1297	1302	1307	1312	1318	1323
9	1328	1334	1339	1344	1350	1355	1360	1365	1371	1376
820	91381	91387	91392	91397	91403	91408	91413	91418	91424	91429
1	1434	1440	1445	1450	1455	1461	1466	1471	1477	1482
2	1487	1492	1498	1503	1508	1514	1519	1524	1529	1535
3	1540	1545	1551	1556	1561	1566	1572	1577	1582	1587
4	1593	1598	1603	1609	1614	1619	1624	1630	1635	1640
5	1645	1651	1656	1661	1666	1672	1677	1682	1687	1693
6	1698	1703	1709	1714	1719	1724	1730	1735	1740	1745
7	1751	1756	1761	1766	1772	1777	1782	1787	1793	1798
8	1803	1808	1814	1819	1824	1829	1834	1840	1845	1850
9	1855	1861	1866	1871	1876	1882	1887	1892	1897	1903
830	91908	91913	91918	91924	91929	91934	91939	91944	91950	91955
1	1960	1965	1971	1976	1981	1986	1991	1997	2002	2007
2	2012	2018	2023	2028	2033	2038	2044	2049	2054	2059
3	2065	2070	2075	2080	2085	2091	2096	2101	2106	2111
4	2117	2122	2127	2132	2137	2143	2148	2153	2158	2163
5	2169	2174	2179	2184	2189	2195	2200	2205	2210	2215
6	2221	2226	2231	2236	2241	2247	2252	2257	2262	2267
7	2273	2278	2283	2288	2293	2298	2304	2309	2314	2319
8	2324	2330	2335	2340	2345	2350	2355	2361	2366	2371
9	2376	2381	2387	2392	2397	2402	2407	2412	2418	2423
840	92428	92433	92438	92443	92449	92454	92459	92464	92469	92474
1	2480	2485	2490	2495	2500	2505	2511	2516	2521	2526
2	2531	2536	2542	2547	2552	2557	2562	2567	2572	2578
3	2583	2588	2593	2598	2603	2609	2614	2619	2624	2629
4	2634	2639	2645	2650	2655	2660	2665	2670	2675	2681
5	2686	2691	2696	2701	2706	2711	2716	2722	2727	2732
6	2737	2742	2747	2752	2758	2763	2768	2773	2778	2783
7	2788	2793	2799	2804	2809	2814	2819	2824	2829	2834
8	2840	2845	2850	2855	2860	2865	2870	2875	2881	2886
9	2891	2896	2901	2906	2911	2916	2921	2927	2932	2937
850	92942	92947	92952	92957	92962	92967	92973	92978	92983	92988

TABLE 1. LOGARITHMS OF NUMBERS

N	0	1	2	3	4	5	6	7	8	9
850	92942	92947	92952	92957	92962	92967	92973	92978	92983	92988
1	2993	2998	3003	3008	3013	3018	3024	3029	3034	3039
2	3044	3049	3054	3059	3064	3069	3075	3080	3085	3090
3	3095	3100	3105	3110	3115	3120	3125	3131	3136	3141
4	3146	3151	3156	3161	3166	3171	3176	3181	3186	3192
5	3197	3202	3207	3212	3217	3222	3227	3232	3237	3242
6	3247	3252	3258	3263	3268	3273	3278	3283	3288	3293
7	3298	3303	3308	3313	3318	3323	3328	3334	3339	3344
8	3349	3354	3359	3364	3369	3374	3379	3384	3389	3394
9	3399	3404	3409	3414	3420	3425	3430	3435	3440	3445
860	93450	93455	93460	93465	93470	93475	93480	93485	93490	93495
1	3500	3505	3510	3515	3520	3526	3531	3536	3541	3546
2	3551	3556	3561	3566	3571	3576	3581	3586	3591	3596
3	3601	3606	3611	3616	3621	3626	3631	3636	3641	3646
4	3651	3656	3661	3666	3671	3676	3682	3687	3692	3697
5	3702	3707	3712	3717	3722	3727	3732	3737	3742	3747
6	3752	3757	3762	3767	3772	3777	3782	3787	3792	3797
7	3802	3807	3812	3817	3822	3827	3832	3837	3842	3847
8	3852	3857	3862	3867	3872	3877	3882	3887	3892	3897
9	3902	3907	3912	3917	3922	3927	3932	3937	3942	3947
870	93952	93957	93962	93967	93972	93977	93982	93987	93992	93997
1	4002	4007	4012	4017	4022	4027	4032	4037	4042	4047
2	4052	4057	4062	4067	4072	4077	4082	4086	4091	4096
3	4101	4106	4111	4116	4121	4126	4131	4136	4141	4146
4	4151	4156	4161	4166	4171	4176	4181	4186	4191	4196
5	4201	4206	4211	4216	4221	4226	4231	4236	4240	4245
6	4250	4255	4260	4265	4270	4275	4280	4285	4290	4295
7	4300	4305	4310	4315	4320	4325	4330	4335	4340	4345
8	4349	4354	4359	4364	4369	4374	4379	4384	4389	4394
9	4399	4404	4409	4414	4419	4424	4429	4433	4438	4443
880	94448	94453	94458	94463	94468	94473	94478	94483	94488	94493
1	4498	4503	4507	4512	4517	4522	4527	4532	4537	4542
2	4547	4552	4557	4562	4567	4571	4576	4581	4586	4591
3	4596	4601	4606	4611	4616	4621	4626	4630	4635	4640
4	4645	4650	4655	4660	4665	4670	4675	4680	4685	4689
5	4694	4699	4704	4709	4714	4719	4724	4729	4734	4738
6	4743	4748	4753	4758	4763	4768	4773	4778	4783	4787
7	4792	4797	4802	4807	4812	4817	4822	4827	4832	4836
8	4841	4846	4851	4856	4861	4866	4871	4876	4880	4885
9	4890	4895	4900	4905	4910	4915	4919	4924	4929	4934
890	94939	94944	94949	94954	94959	94963	94968	94973	94978	94983
1	4988	4993	4998	5002	5007	5012	5017	5022	5027	5032
2	5036	5041	5046	5051	5056	5061	5066	5071	5075	5080
3	5085	5090	5095	5100	5105	5109	5114	5119	5124	5129
4	5134	5139	5143	5148	5153	5158	5163	5168	5173	5177
5	5182	5187	5192	5197	5202	5207	5211	5216	5221	5226
6	5231	5236	5240	5245	5250	5255	5260	5265	5270	5274
7	5279	5284	5289	5294	5299	5303	5308	5313	5318	5323
8	5328	5332	5337	5342	5347	5352	5357	5361	5366	5371
9	5376	5381	5386	5390	5395	5400	5405	5410	5415	5419
900	95424	95429	95434	95439	95444	95448	95453	95458	95463	95468

TABLE 1. LOGARITHMS OF NUMBERS

N	0	1	2	3	4	5	6	7	8	9
900	95424	95429	95434	95439	95444	95448	95453	95458	95463	95468
1	5472	5477	5482	5487	5492	5497	5501	5506	5511	5516
2	5521	5525	5530	5535	5540	5545	5550	5554	5559	5564
3	5569	5574	5578	5583	5588	5593	5598	5602	5607	5612
4	5617	5622	5626	5631	5636	5641	5646	5650	5655	5660
5	5665	5670	5674	5679	5684	5689	5694	5698	5703	5708
6	5713	5718	5722	5727	5732	5737	5742	5746	5751	5756
7	5761	5766	5770	5775	5780	5785	5789	5794	5799	5804
8	5809	5813	5818	5823	5828	5832	5837	5842	5847	5852
9	5856	5861	5866	5871	5875	5880	5885	5890	5895	5899
910	95904	95909	95914	95918	95923	95928	95933	95938	95942	95947
1	5952	5957	5961	5966	5971	5976	5980	5985	5990	5995
2	5999	6004	6009	6014	6019	6023	6028	6033	6038	6042
3	6047	6052	6057	6061	6066	6071	6076	6080	6085	6090
4	6095	6099	6104	6109	6114	6118	6123	6128	6133	6137
5	6142	6147	6152	6156	6161	6166	6171	6175	6180	6185
6	6190	6194	6199	6204	6209	6213	6218	6223	6227	6232
7	6237	6242	6246	6251	6256	6261	6265	6270	6275	6280
8	6284	6289	6294	6298	6303	6308	6313	6317	6322	6327
9	6332	6336	6341	6346	6350	6355	6360	6365	6369	6374
920	96379	96384	96388	96393	96398	96402	96407	96412	96417	96421
1	6426	6431	6435	6440	6445	6450	6454	6459	6464	6468
2	6473	6478	6483	6487	6492	6497	6501	6506	6511	6515
3	6520	6525	6530	6534	6539	6544	6548	6553	6558	6562
4	6567	6572	6577	6581	6586	6591	6595	6600	6605	6609
5	6614	6619	6624	6628	6633	6638	6642	6647	6652	6656
6	6661	6666	6670	6675	6680	6685	6689	6694	6699	6703
7	6708	6713	6717	6722	6727	6731	6736	6741	6745	6750
8	6755	6759	6764	6769	6774	6778	6783	6788	6792	6797
9	6802	6806	6811	6816	6820	6825	6830	6834	6839	6844
930	96848	96853	96858	96862	96867	96872	96876	96881	96886	96890
1	6895	6900	6904	6909	6914	6918	6923	6928	6932	6937
2	6942	6946	6951	6956	6960	6965	6970	6974	6979	6984
3	6988	6993	6997	7002	7007	7011	7016	7021	7025	7030
4	7035	7039	7044	7049	7053	7058	7063	7067	7072	7077
5	7081	7086	7090	7095	7100	7104	7109	7114	7118	7123
6	7128	7132	7137	7142	7146	7151	7155	7160	7165	7169
7	7174	7179	7183	7188	7192	7197	7202	7206	7211	7216
8	7220	7225	7230	7234	7239	7243	7248	7253	7257	7262
9	7267	7271	7276	7280	7285	7290	7294	7299	7304	7308
940	97313	97317	97322	97327	97331	97336	97340	97345	97350	97354
1	7359	7364	7368	7373	7377	7382	7387	7391	7396	7400
2	7405	7410	7414	7419	7424	7428	7433	7437	7442	7447
3	7451	7456	7460	7465	7470	7474	7479	7483	7488	7493
4	7497	7502	7506	7511	7516	7520	7525	7529	7534	7539
5	7543	7548	7552	7557	7562	7566	7571	7575	7580	7585
6	7589	7594	7598	7603	7607	7612	7617	7621	7626	7630
7	7635	7640	7644	7649	7653	7658	7663	7667	7672	7676
8	7681	7685	7690	7695	7699	7704	7708	7713	7717	7722
9	7727	7731	7736	7740	7745	7749	7754	7759	7763	7768
950	97772	97777	97782	97786	97791	97795	97800	97804	97809	97813

TABLE 1. LOGARITHMS OF NUMBERS

N	0	1	2	3	4	5	6	7	8	9
950	97772	97777	97782	97786	97791	97795	97800	97804	97809	97813
1	7818	7823	7827	7832	7836	7841	7845	7850	7855	7859
2	7864	7868	7873	7877	7882	7886	7891	7896	7900	7905
3	7909	7914	7918	7923	7928	7932	7937	7941	7946	7950
4	7955	7959	7964	7968	7973	7978	7982	7987	7991	7996
5	8000	8005	8009	8014	8019	8023	8028	8032	8037	8041
6	8046	8050	8055	8059	8064	8068	8073	8078	8082	8087
7	8091	8096	8100	8105	8109	8114	8118	8123	8127	8132
8	8137	8141	8146	8150	8155	8159	8164	8168	8173	8177
9	8182	8186	8191	8195	8200	8204	8209	8214	8218	8223
960	98227	98232	98236	98241	98245	98250	98254	98259	98263	98268
1	8272	8277	8281	8286	8290	8295	8299	8304	8308	8313
2	8318	8322	8327	8331	8336	8340	8345	8349	8354	8358
3	8363	8367	8372	8376	8381	8385	8390	8394	8399	8403
4	8408	8412	8417	8421	8426	8430	8435	8439	8444	8448
5	8453	8457	8462	8466	8471	8475	8480	8484	8489	8493
6	8498	8502	8507	8511	8516	8520	8525	8529	8534	8538
7	8543	8547	8552	8556	8561	8565	8570	8574	8579	8583
8	8588	8592	8597	8601	8605	8610	8614	8619	8623	8628
9	8632	8637	8641	8646	8650	8655	8659	8664	8668	8673
970	98677	98682	98686	98691	98695	98700	98704	98709	98713	98717
1	8722	8726	8731	8735	8740	8744	8749	8753	8758	8762
2	8767	8771	8776	8780	8784	8789	8793	8798	8802	8807
3	8811	8816	8820	8825	8829	8834	8838	8843	8847	8851
4	8856	8860	8865	8869	8874	8878	8883	8887	8892	8896
5	8900	8905	8909	8914	8918	8923	8927	8932	8936	8941
6	8945	8949	8954	8958	8963	8967	8972	8976	8981	8985
7	8989	8994	8998	9003	9007	9012	9016	9021	9025	9029
8	9034	9038	9043	9047	9052	9056	9061	9065	9069	9074
9	9078	9083	9087	9092	9096	9100	9105	9109	9114	9118
980	99123	99127	99131	99136	99140	99145	99149	99154	99158	99162
1	9167	9171	9176	9180	9185	9189	9193	9198	9202	9207
2	9211	9216	9220	9224	9229	9233	9238	9242	9247	9251
3	9255	9260	9264	9269	9273	9277	9282	9286	9291	9295
4	9300	9304	9308	9313	9317	9322	9326	9330	9335	9339
5	9344	9348	9352	9357	9361	9366	9370	9374	9379	9383
6	9388	9392	9396	9401	9405	9410	9414	9419	9423	9427
7	9432	9436	9441	9445	9449	9454	9458	9463	9467	9471
8	9476	9480	9484	9489	9493	9498	9502	9506	9511	9515
9	9520	9524	9528	9533	9537	9542	9546	9550	9555	9559
990	99564	99568	99572	99577	99581	99585	99590	99594	99599	99603
1	9607	9612	9616	9621	9625	9629	9634	9638	9642	9647
2	9651	9656	9660	9664	9669	9673	9677	9682	9686	9691
3	9695	9699	9704	9708	9712	9717	9721	9726	9730	9734
4	9739	9743	9747	9752	9756	9760	9765	9769	9774	9778
5	9782	9787	9791	9795	9800	9804	9808	9813	9817	9822
6	9826	9830	9835	9839	9843	9848	9852	9856	9861	9865
7	9870	9874	9878	9883	9887	9891	9896	9900	9904	9909
8	9913	9917	9922	9926	9930	9935	9939	9944	9948	9952
9	9957	9961	9965	9970	9974	9978	9983	9987	9991	9996
1000	00000	00004	00009	00013	00017	00022	00026	00030	00035	00039

TABLE 2. LOGARITHMIC SINES AND COSINES

′	0°		1°		2°		′
	Sine	Cosine	Sine	Cosine	Sine	Cosine	
0	—∞	10.00000	8.24186	9.99993	8.54282	9.99974	60
1	6.46373	00000	24903	99993	54642	99973	59
2	76476	00000	25609	99993	54999	99973	58
3	94085	00000	26304	99993	55354	99972	57
4	7.06579	00000	26988	99992	55705	99972	56
5	16270	00000	27661	99992	56054	99971	55
6	24188	00000	28324	99992	56400	99971	54
7	30882	00000	28977	99992	56743	99970	53
8	36682	00000	29621	99992	57084	99970	52
9	41797	00000	30255	99991	57421	99969	51
10	7.46373	10.00000	8.30879	9.99991	8.57757	9.99969	50
11	50512	00000	31495	99991	58089	99968	49
12	54291	00000	32103	99990	58419	99968	48
13	57767	00000	32702	99990	58747	99967	47
14	60985	00000	33292	99990	59072	99967	46
15	63982	00000	33875	99990	59395	99967	45
16	66784	00000	34450	99989	59715	99966	44
17	69417	9.99999	35018	99989	60033	99966	43
18	71900	99999	35578	99989	60349	99965	42
19	74248	99999	36131	99989	60662	99964	41
20	7.76475	9.99999	8.36678	9.99988	8.60973	9.99964	40
21	78594	99999	37217	99988	61282	99963	39
22	80615	99999	37750	99988	61589	99963	38
23	82545	99999	38276	99987	61894	99962	37
24	84393	99999	38796	99987	62196	99962	36
25	86166	99999	39310	99987	62497	99961	35
26	87870	99999	39818	99986	62795	99961	34
27	89509	99999	40320	99986	63091	99960	33
28	91088	99999	40816	99986	63385	99960	32
29	92612	99998	41307	99985	63678	99959	31
30	7.94084	9.99998	8.41792	9.99985	8.63968	9.99959	30
31	95508	99998	42272	99985	64256	99958	29
32	96887	99998	42746	99984	64543	99958	28
33	98223	99998	43216	99984	64827	99957	27
34	99520	99998	43680	99984	65110	99956	26
35	8.00779	99998	44139	99983	65391	99956	25
36	02002	99998	44594	99983	65670	99955	24
37	03192	99997	45044	99983	65947	99955	23
38	04350	99997	45489	99982	66223	99954	22
39	05478	99997	45930	99982	66497	99954	21
40	8.06578	9.99997	8.46366	9.99982	8.66769	9.99953	20
41	07650	99997	46799	99981	67039	99952	19
42	08696	99997	47226	99981	67308	99952	18
43	09718	99997	47650	99981	67575	99951	17
44	10717	99996	48069	99980	67841	99951	16
45	11693	99996	48485	99980	68104	99950	15
46	12647	99996	48896	99979	68367	99949	14
47	13581	99996	49304	99979	68627	99949	13
48	14495	99996	49708	99979	68886	99948	12
49	15391	99996	50108	99978	69144	99948	11
50	8.16268	9.99995	8.50504	9.99978	8.69400	9.99947	10
51	17128	99995	50897	99977	69654	99946	9
52	17971	99995	51287	99977	69907	99946	8
53	18798	99995	51673	99977	70159	99945	7
54	19610	99995	52055	99976	70409	99944	6
55	20407	99994	52434	99976	70658	99944	5
56	21189	99994	52810	99975	70905	99943	4
57	21958	99994	53183	99975	71151	99942	3
58	22713	99994	53552	99974	71395	99942	2
59	23456	99994	53919	99974	71638	99941	1
60	24186	99993	54282	99974	71880	99940	0
′	Cosine	Sine	Cosine	Sine	Cosine	Sine	′
	89°		88°		87°		

TABLE 2. LOGARITHMIC SINES AND COSINES

′	3° Sine	3° Cosine	4° Sine	4° Cosine	5° Sine	5° Cosine	′
0	8.71880	9.99940	8.84358	9.99894	8.94030	9.99834	60
1	72120	99940	84539	99893	94174	99833	59
2	72359	99939	84718	99892	94317	99832	58
3	72597	99938	84897	99891	94461	99831	57
4	72834	99938	85075	99891	94603	99830	56
5	73069	99937	85252	99890	94746	99829	55
6	73303	99936	85429	99889	94887	99828	54
7	73535	99936	85605	99888	95029	99827	53
8	73767	99935	85780	99887	95170	99825	52
9	73997	99934	85955	99886	95310	99824	51
10	8.74226	9.99934	8.86128	9.99885	8.95450	9.99823	50
11	74454	99933	86301	99884	95589	99822	49
12	74680	99932	86474	99883	95728	99821	48
13	74906	99932	86645	99882	95867	99820	47
14	75130	99931	86816	99881	96005	99819	46
15	75353	99930	86987	99880	96143	99817	45
16	75575	99929	87156	99879	96280	99816	44
17	75795	99929	87325	99879	96417	99815	43
18	76015	99928	87494	99878	96553	99814	42
19	76234	99927	87661	99877	96689	99813	41
20	8.76451	9.99926	8.87829	9.99876	8.96825	9.99812	40
21	76667	99926	87995	99875	96960	99810	39
22	76883	99925	88161	99874	97095	99809	38
23	77097	99924	88326	99873	97229	99808	37
24	77310	99923	88490	99872	97363	99807	36
25	77522	99923	88654	99871	97496	99806	35
26	77733	99922	88817	99870	97629	99804	34
27	77943	99921	88980	99869	97762	99803	33
28	78152	99920	89142	99868	97894	99802	32
29	78360	99920	89304	99867	98026	99801	31
30	8.78568	9.99919	8.89464	9.99866	8.98157	9.99800	30
31	78774	99918	89625	99865	98288	99798	29
32	78979	99917	89784	99864	98419	99797	28
33	79183	99917	89943	99863	98549	99796	27
34	79386	99916	90102	99862	98679	99795	26
35	79588	99915	90260	99861	98808	99793	25
36	79789	99914	90417	99860	98937	99792	24
37	79990	99913	90574	99859	99066	99791	23
38	80189	99913	90730	99858	99194	99790	22
39	80388	99912	90885	99857	99322	99788	21
40	8.80585	9.99911	8.91040	9.99856	8.99450	9.99787	20
41	80782	99910	91195	99855	99577	99786	19
42	80978	99909	91349	99854	99704	99785	18
43	81173	99909	91502	99853	99830	99783	17
44	81367	99908	91655	99852	99956	99782	16
45	81560	99907	91807	99851	9.00082	99781	15
46	81752	99906	91959	99850	00207	99780	14
47	81944	99905	92110	99848	00332	99778	13
48	82134	99904	92261	99847	00456	99777	12
49	82324	99904	92411	99846	00581	99776	11
50	8.82513	9.99903	8.92561	9.99845	9.00704	9.99775	10
51	82701	99902	92710	99844	00828	99773	9
52	82888	99901	92859	99843	00951	99772	8
53	83075	99900	93007	99842	01074	99771	7
54	83261	99899	93154	99841	01196	99769	6
55	83446	99898	93301	99840	01318	99768	5
56	83630	99898	93448	99839	01440	99767	4
57	83813	99897	93594	99838	01561	99765	3
58	83996	99896	93740	99837	01682	99764	2
59	84177	99895	93885	99836	01803	99763	1
60	84358	99894	94030	99834	01923	99761	0
′	Cosine	Sine	Cosine	Sine	Cosine	Sine	′
	86°		85°		84°		

215

TABLE 2. LOGARITHMIC SINES AND COSINES

′	6°		7°		8°		′
	Sine	Cosine	Sine	Cosine	Sine	Cosine	
0	9.01923	9.99761	9.08589	9.99675	9.14356	9.99575	60
1	02043	99760	08692	99674	14445	99574	59
2	02163	99759	08795	99672	14535	99572	58
3	02283	99757	08897	99670	14624	99570	57
4	02402	99756	08999	99669	14714	99568	56
5	02520	99755	09101	99667	14803	99566	55
6	02639	99753	09202	99666	14891	99565	54
7	02757	99752	09304	99664	14980	99563	53
8	02874	99751	09405	99663	15069	99561	52
9	02992	99749	09506	99661	15157	99559	51
10	9.03109	9.99748	9.09606	9.99659	9.15245	9.99557	50
11	03226	99747	09707	99658	15333	99556	49
12	03342	99745	09807	99656	15421	99554	48
13	03458	99744	09907	99655	15508	99552	47
14	03574	99742	10006	99653	15596	99550	46
15	03690	99741	10106	99651	15683	99548	45
16	03805	99740	10205	99650	15770	99546	44
17	03920	99738	10304	99648	15857	99545	43
18	04034	99737	10402	99647	15944	99543	42
19	04149	99736	10501	99645	16030	99541	41
20	9.04262	9.99734	9.10599	9.99643	9.16116	9.99539	40
21	04376	99733	10697	99642	16203	99537	39
22	04490	99731	10795	99640	16289	99535	38
23	04603	99730	10893	99638	16374	99533	37
24	04715	99728	10990	99637	16460	99532	36
25	04828	99727	11087	99635	16545	99530	35
26	04940	99726	11184	99633	16631	99528	34
27	05052	99724	11281	99632	16716	99526	33
28	05164	99723	11377	99630	16801	99524	32
29	05275	99721	11474	99629	16886	99522	31
30	9.05386	9.99720	9.11570	9.99627	9.16970	9.99520	30
31	05497	99718	11666	99625	17055	99518	29
32	05607	99717	11761	99624	17139	99517	28
33	05717	99716	11857	99622	17223	99515	27
34	05827	99714	11952	99620	17307	99513	26
35	05937	99713	12047	99618	17391	99511	25
36	06046	99711	12142	99617	17474	99509	24
37	06155	99710	12236	99615	17558	99507	23
38	06264	99708	12331	99613	17641	99505	22
39	06372	99707	12425	99612	17724	99503	21
40	9.06481	9.99705	9.12519	9.99610	9.17807	9.99501	20
41	06589	99704	12612	99608	17890	99499	19
42	06696	99702	12706	99607	17973	99497	18
43	06804	99701	12799	99605	18055	99495	17
44	06911	99699	12892	99603	18137	99494	16
45	07018	99698	12985	99601	18220	99492	15
46	07124	99696	13078	99600	18302	99490	14
47	07231	99695	13171	99598	18383	99488	13
48	07337	99693	13263	99596	18465	99486	12
49	07442	99692	13355	99595	18547	99484	11
50	9.07548	9.99690	9.13447	9.99593	9.18628	9.99482	10
51	07653	99689	13539	99591	18709	99480	9
52	07758	99687	13630	99589	18790	99478	8
53	07863	99686	13722	99588	18871	99476	7
54	07968	99684	13813	99586	18952	99474	6
55	08072	99683	13904	99584	19033	99472	5
56	08176	99681	13994	99582	19113	99470	4
57	08280	99680	14085	99581	19193	99468	3
58	08383	99678	14175	99579	19273	99466	2
59	08486	99677	14266	99577	19353	99464	1
60	08589	99675	14356	99575	19433	99462	0
′	Cosine	Sine	Cosine	Sine	Cosine	Sine	′
	83°		82°		81°		

TABLE 2. LOGARITHMIC SINES AND COSINES

′	9°		10°		11°		′
	Sine	Cosine	Sine	Cosine	Sine	Cosine	
0	9.19433	9.99462	9.23967	9.99335	9.28060	9.99195	60
1	19513	99460	24039	99333	28125	99192	59
2	19592	99458	24110	99331	28190	99190	58
3	19672	99456	24181	99328	28254	99187	57
4	19751	99454	24253	99326	28319	99185	56
5	19830	99452	24324	99324	28384	99182	55
6	19909	99450	24395	99322	28448	99180	54
7	19988	99448	24466	99319	28512	99177	53
8	20067	99446	24536	99317	28577	99175	52
9	20145	99444	24607	99315	28641	99172	51
10	9.20223	9.99442	9.24677	9.99313	9.28705	9.99170	50
11	20302	99440	24748	99310	28769	99167	49
12	20380	99438	24818	99308	28833	99165	48
13	20458	99436	24888	99306	28896	99162	47
14	20535	99434	24958	99304	28960	99160	46
15	20613	99432	25028	99301	29024	99157	45
16	20691	99429	25098	99299	29087	99155	44
17	20768	99427	25168	99297	29150	99152	43
18	20845	99425	25237	99294	29214	99150	42
19	20922	99423	25307	99292	29277	99147	41
20	9.20999	9.99421	9.25376	9.99290	9.29340	9.99145	40
21	21076	99419	25445	99288	29403	99142	39
22	21153	99417	25514	99285	29466	99140	38
23	21229	99415	25583	99283	29529	99137	37
24	21306	99413	25652	99281	29591	99135	36
25	21382	99411	25721	99278	29654	99132	35
26	21458	99409	25790	99276	29716	99130	34
27	21534	99407	25858	99274	29779	99127	33
28	21610	99404	25927	99271	29841	99124	32
29	21685	99402	25995	99269	29903	99122	31
30	9.21761	9.99400	9.26063	9.99267	9.29966	9.99119	30
31	21836	99398	26131	99264	30028	99117	29
32	21912	99396	26199	99262	30090	99114	28
33	21987	99394	26267	99260	30151	99112	27
34	22062	99392	26335	99257	30213	99109	26
35	22137	99390	26403	99255	30275	99106	25
36	22211	99388	26470	99252	30336	99104	24
37	22286	99385	26538	99250	30398	99101	23
38	22361	99383	26605	99248	30459	99099	22
39	22435	99381	26672	99245	30521	99096	21
40	9.22509	9.99379	9.26739	9.99243	9.30582	9.99093	20
41	22583	99377	26806	99241	30643	99091	19
42	22657	99375	26873	99238	30704	99088	18
43	22731	99372	26940	99236	30765	99086	17
44	22805	99370	27007	99233	30826	99083	16
45	22878	99368	27073	99231	30887	99080	15
46	22952	99366	27140	99229	30947	99078	14
47	23025	99364	27206	99226	31008	99075	13
48	23098	99362	27273	99224	31068	99072	12
49	23171	99359	27339	99221	31129	99070	11
50	9.23244	9.99357	9.27405	9.99219	9.31189	9.99067	10
51	23317	99355	27471	99217	31250	99064	9
52	23390	99353	27537	99214	31310	99062	8
53	23462	99351	27602	99212	31370	99059	7
54	23535	99348	27668	99209	31430	99056	6
55	23607	99346	27734	99207	31490	99054	5
56	23679	99344	27799	99204	31549	99051	4
57	23752	99342	27864	99202	31609	99048	3
58	23823	99340	27930	99200	31669	99046	2
59	23895	99337	27995	99197	31728	99043	1
60	23967	99335	28060	99195	31788	99040	0
′	Cosine	Sine	Cosine	Sine	Cosine	Sine	′
	80°		79°		78°		

TABLE 2. LOGARITHMIC SINES AND COSINES

′	12°		13°		14°		′
	Sine	Cosine	Sine	Cosine	Sine	Cosine	
0	9.31788	9.99040	9.35209	9.98872	9.38368	9.98690	60
1	31847	99038	35263	98869	38418	98687	59
2	31907	99035	35318	98867	38469	98684	58
3	31966	99032	35373	98864	38519	98681	57
4	32025	99030	35427	98861	38570	98678	56
5	32084	99027	35481	98858	38620	98675	55
6	32143	99024	35536	98855	38670	98671	54
7	32202	99022	35590	98852	38721	98668	53
8	32261	99019	35644	98849	38771	98665	52
9	32319	99016	35698	98846	38821	98662	51
10	9.32378	9.99013	9.35752	9.98843	9.38871	9.98659	50
11	32437	99011	35806	98840	38921	98656	49
12	32495	99008	35860	98837	38971	98652	48
13	32553	99005	35914	98834	39021	98649	47
14	32612	99002	35968	98831	39071	98646	46
15	32670	99000	36022	98828	39121	98643	45
16	32728	98997	36075	98825	39170	98640	44
17	32786	98994	36129	98822	39220	98636	43
18	32844	98991	36182	98819	39270	98633	42
19	32902	98989	36236	98816	39319	98630	41
20	9.32960	9.98986	9.36289	9.98813	9.39369	9.98627	40
21	33018	98983	36342	98810	39418	98623	39
22	33075	98980	36395	98807	39467	98620	38
23	33133	98978	36449	98804	39517	98617	37
24	33190	98975	36502	98801	39566	98614	36
25	33248	98972	36555	98798	39615	98610	35
26	33305	98969	36608	98795	39664	98607	34
27	33362	98967	36660	98792	39715	98604	33
28	33420	98964	36713	98789	39762	98601	32
29	33477	98961	36766	98786	39811	98597	31
30	9.33534	9.98958	9.36819	9.98783	9.39860	9.98594	30
31	33591	98955	36871	98780	39909	98591	29
32	33647	98953	36924	98777	39958	98588	28
33	33704	98950	36976	98774	40006	98584	27
34	33761	98947	37028	98771	40055	98581	26
35	33818	98944	37081	98768	40103	98578	25
36	33874	98941	37133	98765	40152	98574	24
37	33931	98938	37185	98762	40200	98571	23
38	33987	98936	37237	98759	40249	98568	22
39	34043	98933	37289	98756	40297	98565	21
40	9.34100	9.98930	9.37341	9.98753	9.40346	9.98561	20
41	34156	98927	37393	98750	40394	98558	19
42	34212	98924	37445	98746	40442	98555	18
43	34268	98921	37497	98743	40490	98551	17
44	34324	98919	37549	98740	40538	98548	16
45	34380	98916	37600	98737	40586	98545	15
46	34436	98913	37652	98734	40634	98541	14
47	34491	98910	37703	98731	40682	98538	13
48	34547	98907	37755	98728	40730	98535	12
49	34602	98904	37806	98725	40778	98531	11
50	9.34658	9.98901	9.37858	9.98722	9.40825	9.98528	10
51	34713	98898	37909	98719	40873	98525	9
52	34769	98896	37960	98715	40921	98521	8
53	34824	98893	38011	98712	40968	98518	7
54	34879	98890	38062	98709	41016	98515	6
55	34934	98887	38113	98706	41063	98511	5
56	34989	98884	38164	98703	41111	98508	4
57	35044	98881	38215	98700	41158	98505	3
58	35099	98878	38266	98697	41205	98501	2
59	35154	98875	38317	98694	41252	98498	1
60	35209	98872	38368	98690	41300	98494	0
′	Cosine	Sine	Cosine	Sine	Cosine	Sine	′
	77°		76°		75°		

TABLE 2. LOGARITHMIC SINES AND COSINES

′	15° Sine	15° Cosine	16° Sine	16° Cosine	17° Sine	17° Cosine	′
0	9.41300	9.98494	9.44034	9.98284	9.46594	9.98060	60
1	41347	98491	44078	98281	46635	98056	59
2	41394	98488	44122	98277	46676	98052	58
3	41441	98484	44166	98273	46717	98048	57
4	41488	98481	44210	98270	46758	98044	56
5	41535	98477	44253	98266	46800	98040	55
6	41582	98474	44297	98262	46841	98036	54
7	41628	98471	44341	98259	46882	98032	53
8	41675	98467	44385	98255	46923	98029	52
9	41722	98464	44428	98251	46964	98025	51
10	9.41768	9.98460	9.44472	9.98248	9.47005	9.98021	50
11	41815	98457	44516	98244	47045	98017	49
12	41861	98453	44559	98240	47086	98013	48
13	41908	98450	44602	98237	47127	98009	47
14	41954	98447	44646	98233	47168	98005	46
15	42001	98443	44689	98229	47209	98001	45
16	42047	98440	44733	98226	47249	97997	44
17	42093	98436	44776	98222	47290	97993	43
18	42140	98433	44819	98218	47330	97989	42
19	42186	98429	44862	98215	47371	97986	41
20	9.42232	9.98426	9.44905	9.98211	9.47411	9.97982	40
21	42278	98422	44948	98207	47452	97978	39
22	42324	98419	44992	98204	47492	97974	38
23	42370	98415	45035	98200	47533	97970	37
24	42416	98412	45077	98196	47573	97966	36
25	42461	98409	45120	98192	47613	97962	35
26	42507	98405	45163	98189	47654	97958	34
27	42553	98402	45206	98185	47694	97954	33
28	42599	98398	45249	98181	47734	97950	32
29	42644	98395	45292	98177	47774	97946	31
30	9.42690	9.98391	9.45334	9.98174	9.47814	9.97942	30
31	42735	98388	45377	98170	47854	97938	29
32	42781	98384	45419	98166	47894	97934	28
33	42826	98381	45462	98162	47934	97930	27
34	42872	98377	45504	98159	47974	97926	26
35	42917	98373	45547	98155	48014	97922	25
36	42962	98370	45589	98151	48054	97918	24
37	43008	98366	45632	98147	48094	97914	23
38	43053	98363	45674	98144	48133	97910	22
39	43098	98359	45716	98140	48173	97906	21
40	9.43143	9.98356	9.45758	9.98136	9.48213	9.97902	20
41	43188	98352	45801	98132	48252	97898	19
42	43233	98349	45843	98129	48292	97894	18
43	43278	98345	45885	98125	48332	97890	17
44	43323	98342	45927	98121	48371	97886	16
45	43367	98338	45969	98117	48411	97882	15
46	43412	98334	46011	98113	48450	97878	14
47	43457	98331	46053	98110	48490	97874	13
48	43502	98327	46095	98106	48529	97870	12
49	43546	98324	46136	98102	48568	97866	11
50	9.43591	9.98320	9.46178	9.98098	9.48607	9.97861	10
51	43635	98317	46220	98094	48647	97857	9
52	43680	98313	46262	98090	48686	97853	8
53	43724	98309	46303	98087	48725	97849	7
54	43769	98306	46345	98083	48764	97845	6
55	43813	98302	46386	98079	48803	97841	5
56	43857	98299	46428	98075	48842	97837	4
57	43901	98295	46469	98071	48881	97833	3
58	43946	98291	46511	98067	48920	97829	2
59	43990	98288	46552	98063	48959	97825	1
60	44034	98284	46594	98060	48998	97821	0
′	Cosine	Sine	Cosine	Sine	Cosine	Sine	′
	74°		73°		72°		

TABLE 2. LOGARITHMIC SINES AND COSINES

′	18°		19°		20°		′
	Sine	Cosine	Sine	Cosine	Sine	Cosine	
0	9.48998	9.97821	9.51264	9.97567	9.53405	9.97299	60
1	49037	97817	51301	97563	53440	97294	59
2	49076	97812	51338	97558	53475	97289	58
3	49115	97808	51374	97554	53509	97285	57
4	49153	97804	51411	97550	53544	97280	56
5	49192	97800	51447	97545	53578	97276	55
6	49231	97796	51484	97541	53613	97271	54
7	49269	97792	51520	97536	53647	97266	53
8	49308	97788	51557	97532	53682	97262	52
9	49347	97784	51593	97528	53716	97257	51
10	9.49385	9.97779	9.51629	9.97523	9.53751	9.97252	50
11	49424	97775	51666	97519	53785	97248	49
12	49462	97771	51702	97515	53819	97243	48
13	49500	97767	51738	97510	53854	97238	47
14	49539	97763	51774	97506	53888	97234	46
15	49577	97759	51811	97501	53922	97229	45
16	49615	97754	51847	97497	53957	97224	44
17	49654	97750	51883	97492	53991	97220	43
18	49692	97746	51919	97488	54025	97215	42
19	49730	97742	51955	97484	54059	97210	41
20	9.49768	9.97738	9.51991	9.97479	9.54093	9.97206	40
21	49806	97734	52027	97475	54127	97201	39
22	49844	97729	52063	97470	54161	97196	38
23	49882	97725	52099	97466	54195	97192	37
24	49920	97721	52135	97461	54229	97187	36
25	49958	97717	52171	97457	54263	97182	35
26	49996	97713	52207	97453	54297	97178	34
27	50034	97708	52242	97448	54331	97173	33
28	50072	97704	52278	97444	54365	97168	32
29	50110	97700	52314	97439	54399	97163	31
30	9.50148	9.97696	9.52350	9.97435	9.54433	9.97159	30
31	50185	97691	52385	97430	54466	97154	29
32	50223	97687	52421	97426	54500	97149	28
33	50261	97683	52456	97421	54534	97145	27
34	50298	97679	52492	97417	54567	97140	26
35	50336	97674	52527	97412	54601	97135	25
36	50374	97670	52563	97408	54635	97130	24
37	50411	97666	52598	97403	54668	97126	23
38	50449	97662	52634	97399	54702	97121	22
39	50486	97657	52669	97394	54735	97116	21
40	9.50523	9.97653	9.52705	9.97390	9.54769	9.97111	20
41	50561	97649	52740	97385	54802	97107	19
42	50598	97645	52775	97381	54836	97102	18
43	50635	97640	52811	97376	54869	97097	17
44	50673	97636	52846	97372	54903	97092	16
45	50710	97632	52881	97367	54936	97087	15
46	50747	97628	52916	97363	54969	97083	14
47	50784	97623	52951	97358	55003	97078	13
48	50821	97619	52986	97353	55036	97073	12
49	50858	97615	53021	97349	55069	97068	11
50	9.50896	9.97610	9.53056	9.97344	9.55102	9.97063	10
51	50933	97606	53092	97340	55136	97059	9
52	50970	97602	53126	97335	55169	97054	8
53	51007	97597	53161	97331	55202	97049	7
54	51043	97593	53196	97326	55235	97044	6
55	51080	97589	53231	97322	55268	97039	5
56	51117	97584	53266	97317	55301	97035	4
57	51154	97580	53301	97312	55334	97030	3
58	51191	97576	53336	97308	55367	97025	2
59	51227	97571	53370	97303	55400	97020	1
60	51264	97567	53405	97299	55433	97015	0
′	Cosine	Sine	Cosine	Sine	Cosine	Sine	′
	71°		70°		69°		

TABLE 2. LOGARITHMIC SINES AND COSINES

′	21° Sine	21° Cosine	22° Sine	22° Cosine	23° Sine	23° Cosine	′
0	9.55433	9.97015	9.57358	9.96717	9.59188	9.96403	60
1	55466	97010	57389	96711	59218	96397	59
2	55499	97005	57420	96706	59247	96392	58
3	55532	97001	57451	96701	59277	96387	57
4	55564	96996	57482	96696	59307	96381	56
5	55597	96991	57514	96691	59336	96376	55
6	55630	96986	57545	96686	59366	96370	54
7	55663	96981	57576	96681	59396	96365	53
8	55695	96976	57607	96676	59425	96360	52
9	55728	96971	57638	96670	59455	96354	51
10	9.55761	9.96966	9.57669	9.96665	9.59484	9.96349	50
11	55793	96962	57700	96660	59514	96343	49
12	55826	96957	57731	96655	59543	96338	48
13	55858	96952	57762	96650	59573	96333	47
14	55891	96947	57793	96645	59602	96327	46
15	55923	96942	57824	96640	59632	96322	45
16	55956	96937	57855	96634	59661	96316	44
17	55988	96932	57885	96629	59690	96311	43
18	56021	96927	57916	96624	59720	96305	42
19	56053	96922	57947	96619	59749	96300	41
20	9.56085	9.96917	9.57978	9.96614	9.59778	9.96294	40
21	56118	96912	58008	96608	59808	96289	39
22	56150	96907	58039	96603	59837	96284	38
23	56182	96903	58070	96598	59866	96278	37
24	56215	96898	58101	96593	59895	96273	36
25	56247	96893	58131	96588	59924	96267	35
26	56279	96888	58162	96582	59954	96262	34
27	56311	96883	58192	96577	59983	96256	33
28	56343	96878	58223	96572	60012	96251	32
29	56375	96873	58253	96567	60041	96245	31
30	9.56408	9.96868	9.58284	9.96562	9.60070	9.96240	30
31	56440	96863	58314	96556	60099	96234	29
32	56472	96858	58345	96551	60128	96229	28
33	56504	96853	58375	96546	60157	96223	27
34	56536	96848	58406	96541	60186	96218	26
35	56568	96843	58436	96535	60215	96212	25
36	56599	96838	58467	96530	60244	96207	24
37	56631	96833	58497	96525	60273	96201	23
38	56663	96828	58527	96520	60302	96196	22
39	56695	96823	58557	96514	60331	96190	21
40	9.56727	9.96818	9.58588	9.96509	9.60359	9.96185	20
41	56759	96813	58618	96504	60388	96179	19
42	56790	96808	58648	96498	60417	96174	18
43	56822	96803	58678	96493	60446	96168	17
44	56854	96798	58709	96488	60474	96162	16
45	56886	96793	58739	96483	60503	96157	15
46	56917	96788	58769	96477	60532	96151	14
47	56949	96783	58799	96472	60561	96146	13
48	56980	96778	58829	96467	60589	96140	12
49	57012	96772	58859	96461	60618	96135	11
50	9.57044	9.96767	9.58889	9.96456	9.60646	9.96129	10
51	57075	96762	58919	96451	60675	96123	9
52	57107	96757	58949	96445	60704	96118	8
53	57138	96752	58979	96440	60732	96112	7
54	57169	96747	59009	96435	60761	96107	6
55	57201	96742	59039	96429	60789	96101	5
56	57232	96737	59069	96424	60818	96095	4
57	57264	96732	59098	96419	60846	96090	3
58	57295	96727	59128	96413	60875	96084	2
59	57326	96722	59158	96408	60903	96079	1
60	57358	96717	59188	96403	60931	96073	0
′	Cosine	Sine	Cosine	Sine	Cosine	Sine	′
	68°		67°		66°		

TABLE 2. LOGARITHMIC SINES AND COSINES

′	24°		25°		26°		′
	Sine	Cosine	Sine	Cosine	Sine	Cosine	
0	9.60931	9.96073	9.62595	9.95728	9.64184	9.95366	60
1	60960	96067	62622	95722	64210	95360	59
2	60988	96062	62649	95716	64236	95354	58
3	61016	96056	62676	95710	64262	95348	57
4	61045	96050	62703	95704	64288	95341	56
5	61073	96045	62730	95698	64313	95335	55
6	61101	96039	62757	95692	64339	95329	54
7	61129	96034	62784	95686	64365	95323	53
8	61158	96028	62811	95680	64391	95317	52
9	61186	96022	62838	95674	64417	95310	51
10	9.61214	9.96017	9.62865	9.95668	9.64442	9.95304	50
11	61242	96011	62892	95663	64468	95298	49
12	61270	96005	62918	95657	64494	95292	48
13	61298	96000	62945	95651	64519	95286	47
14	61326	95994	62972	95645	64545	95279	46
15	61354	95988	62999	95639	64571	95273	45
16	61382	95982	63026	95633	64596	95267	44
17	61411	95977	63052	95627	64622	95261	43
18	61438	95971	63079	95621	64647	95254	42
19	61466	95965	63106	95615	64673	95248	41
20	9.61494	9.95960	9.63133	9.95609	9.64698	9.95242	40
21	61522	95954	63159	95603	64724	95236	39
22	61550	95948	63186	95597	64749	95229	38
23	61578	95942	63213	95591	64775	95223	37
24	61606	95937	63239	95585	64800	95217	36
25	61634	95931	63266	95579	64826	95211	35
26	61662	95925	63292	95573	64851	95204	34
27	61689	95920	63319	95567	64877	95198	33
28	61717	95914	63345	95561	64902	95192	32
29	61745	95908	63372	95555	64927	95185	31
30	9.61773	9.95902	9.63398	9.95549	9.64953	9.95179	30
31	61800	95897	63425	95543	64978	95173	29
32	61828	95891	63451	95537	65003	95167	28
33	61856	95885	63478	95531	65029	95160	27
34	61883	95879	63504	95525	65054	95154	26
35	61911	95873	63531	95519	65079	95148	25
36	61939	95868	63557	95513	65104	95141	24
37	61966	95862	63583	95507	65130	95135	23
38	61994	95856	63610	95500	65155	95129	22
39	62021	95850	63636	95494	65180	95122	21
40	9.62049	9.95844	9.63662	9.95488	9.65205	9.95116	20
41	62076	95839	63689	95482	65230	95110	19
42	62104	95833	63715	95476	65255	95103	18
43	62131	95827	63741	95470	65281	95097	17
44	62159	95821	63767	95464	65306	95090	16
45	62186	95815	63794	95458	65331	95084	15
46	62214	95810	63820	95452	65356	95078	14
47	62241	95804	63846	95446	65381	95071	13
48	62268	95798	63872	95440	65406	95065	12
49	62296	95792	63898	95434	65431	95059	11
50	9.62323	9.95786	9.63924	9.95427	9.65456	9.95052	10
51	62350	95780	63950	95421	65481	95046	9
52	62377	95775	63976	95415	65506	95039	8
53	62405	95769	64002	95409	65531	95033	7
54	62432	95763	64028	95403	65556	95027	6
55	62459	95757	64054	95397	65580	95020	5
56	62486	95751	64080	95391	65605	95014	4
57	62513	95745	64106	95384	65630	95007	3
58	62541	95739	64132	95378	65655	95001	2
59	62568	95733	64158	95372	65680	94995	1
60	62595	95728	64184	95366	65705	94988	0
′	Cosine	Sine	Cosine	Sine	Cosine	Sine	′
	65°		64°		63°		

222

TABLE 2. LOGARITHMIC SINES AND COSINES

′	27° Sine	27° Cosine	28° Sine	28° Cosine	29° Sine	29° Cosine	′
0	9.65705	9.94988	9.67161	9.94593	9.68557	9.94182	60
1	65729	94982	67185	94587	68580	94175	59
2	65754	94975	67208	94580	68603	94168	58
3	65779	94969	67232	94573	68625	94161	57
4	65804	94962	67256	94567	68648	94154	56
5	65828	94956	67280	94560	68671	94147	55
6	65853	94949	67303	94553	68694	94140	54
7	65878	94943	67327	94546	68716	94133	53
8	65902	94936	67350	94540	68739	94126	52
9	65927	94930	67374	94533	68762	94119	51
10	9.65952	9.94923	9.67398	9.94526	9.68784	9.94112	50
11	65976	94917	67421	94519	68807	94105	49
12	66001	94911	67445	94513	68829	94098	48
13	66025	94904	67468	94506	68852	94090	47
14	66050	94898	67492	94499	68875	94083	46
15	66075	94891	67515	94492	68897	94076	45
16	66099	94885	67539	94485	68920	94069	44
17	66124	94878	67562	94479	68942	94062	43
18	66148	94871	67586	94472	68965	94055	42
19	66173	94865	67609	94465	68987	94048	41
20	9.66197	9.94858	9.67633	9.94458	9.69010	9.94041	40
21	66221	94852	67656	94451	69032	94034	39
22	66246	94845	67680	94445	69055	94027	38
23	66270	94839	67703	94438	69077	94020	37
24	66295	94832	67726	94431	69100	94012	36
25	66319	94826	67750	94424	69122	94005	35
26	66343	94819	67773	94417	69144	93998	34
27	66368	94813	67796	94410	69167	93991	33
28	66392	94806	67820	94404	69189	93984	32
29	66416	94799	67843	94397	69212	93977	31
30	9.66441	9.94793	9.67866	9.94390	9.69234	9.93970	30
31	66465	94786	67890	94383	69256	93963	29
32	66489	94780	67913	94376	69279	93955	28
33	66513	94773	67936	94369	69301	93948	27
34	66537	94767	67959	94362	69323	93941	26
35	66562	94760	67982	94355	69345	93934	25
36	66586	94753	68006	94349	69368	93927	24
37	66610	94747	68029	94342	69390	93920	23
38	66634	94740	68052	94335	69412	93912	22
39	66658	94734	68075	94328	69434	93905	21
40	9.66682	9.94727	9.68098	9.94321	9.69456	9.93898	20
41	66706	94720	68121	94314	69479	93891	19
42	66731	94714	68144	94307	69501	93884	18
43	66755	94707	68167	94300	69523	93876	17
44	66779	94700	68190	94293	69545	93869	16
45	66803	94694	68213	94286	69567	93862	15
46	66827	94687	68237	94279	69589	93855	14
47	66851	94680	68260	94273	69611	93847	13
48	66875	94674	68283	94266	69633	93840	12
49	66899	94667	68305	94259	69655	93833	11
50	9.66922	9.94660	9.68328	9.94252	9.69677	9.93826	10
51	66946	94654	68351	94245	69699	93819	9
52	66970	94647	68374	94238	69721	93811	8
53	66994	94640	68397	94231	69743	93804	7
54	67018	94634	68420	94224	69765	93797	6
55	67042	94627	68443	94217	69787	93789	5
56	67066	94620	68466	94210	69809	93782	4
57	67090	94614	68489	94203	69831	93775	3
58	67113	94607	68512	94196	69853	93768	2
59	67137	94600	68534	94189	69875	93760	1
60	67161	94593	68557	94182	69897	93753	0
′	Cosine	Sine	Cosine	Sine	Cosine	Sine	′
	62°		61°		60°		

TABLE 2. LOGARITHMIC SINES AND COSINES

′	30°		31°		32°		′
	Sine	Cosine	Sine	Cosine	Sine	Cosine	
0	9.69897	9.93753	9.71184	9.93307	9.72421	9.92842	60
1	69919	93746	71205	93299	72441	92834	59
2	69941	93738	71226	93291	72461	92826	58
3	69963	93731	71247	93284	72482	92818	57
4	69984	93724	71268	93276	72502	92810	56
5	70006	93717	71289	93269	72522	92803	55
6	70028	93709	71310	93261	72542	92795	54
7	70050	93702	71331	93253	72562	92787	53
8	70072	93695	71352	93246	72582	92779	52
9	70093	93687	71373	93238	72602	92771	51
10	9.70115	9.93680	9.71393	9.93230	9.72622	9.92763	50
11	70137	93673	71414	93223	72643	92755	49
12	70159	93665	71435	93215	72663	92747	48
13	70180	93658	71456	93207	72683	92739	47
14	70202	93650	71477	93200	72703	92731	46
15	70224	93643	71498	93192	72723	92723	45
16	70245	93636	71519	93184	72743	92715	44
17	70267	93628	71539	93177	72763	92707	43
18	70288	93621	71560	93169	72783	92699	42
19	70310	93614	71581	93161	72803	92691	41
20	9.70332	9.93606	9.71602	9.93154	9.72823	9.92683	40
21	70353	93599	71622	93146	72843	92675	39
22	70375	93591	71643	93138	72863	92667	38
23	70396	93584	71664	93131	72883	92659	37
24	70418	93577	71685	93123	72902	92651	36
25	70439	93569	71705	93115	72922	92643	35
26	70461	93562	71726	93108	72942	92635	34
27	70482	93554	71747	93100	72962	92627	33
28	70504	93547	71767	93092	72982	92619	32
29	70525	93539	71788	93084	73002	92611	31
30	9.70547	9.93532	9.71809	9.93077	9.73022	9.92603	30
31	70568	93525	71829	93069	73041	92595	29
32	70590	93517	71850	93061	73061	92587	28
33	70611	93510	71870	93053	73081	92579	27
34	70633	93502	71891	93046	73101	92571	26
35	70654	93495	71911	93038	73121	92563	25
36	70675	93487	71932	93030	73140	92555	24
37	70697	93480	71952	93022	73160	92546	23
38	70718	93472	71973	93014	73180	92538	22
39	70739	93465	71994	93007	73200	92530	21
40	9.70761	9.93457	9.72014	9.92999	9.73219	9.92522	20
41	70782	93450	72034	92991	73239	92514	19
42	70803	93442	72055	92983	73259	92506	18
43	70824	93435	72075	92976	73278	92498	17
44	70846	93427	72096	92968	73298	92490	16
45	70867	93420	72116	92960	73318	92482	15
46	70888	93412	72137	92952	73337	92473	14
47	70909	93405	72157	92944	73357	92465	13
48	70931	93397	72177	92936	73377	92457	12
49	70952	93390	72198	92929	73396	92449	11
50	9.70973	9.93382	9.72218	9.92921	9.73416	9.92441	10
51	70994	93375	72238	92913	73435	92433	9
52	71015	93367	72259	92905	73455	92425	8
53	71036	93360	72279	92897	73474	92416	7
54	71058	93352	72299	92889	73494	92408	6
55	71079	93344	72320	92881	73513	92400	5
56	71100	93337	72340	92874	73533	92392	4
57	71121	93329	72360	92866	73552	92384	3
58	71142	93322	72381	92858	73572	92376	2
59	71163	93314	72401	92850	73591	92367	1
60	71184	93307	72421	92842	73611	92359	0
′	Cosine	Sine	Cosine	Sine	Cosine	Sine	′
	59°		58°		57°		

TABLE 2. LOGARITHMIC SINES AND COSINES

′	33°		34°		35°		′
	Sine	Cosine	Sine	Cosine	Sine	Cosine	
0	9.73611	9.92359	9.74756	9.91857	9.75859	9.91336	60
1	73630	92351	74775	91849	75877	91328	59
2	73650	92343	74794	91840	75895	91319	58
3	73669	92335	74812	91832	75913	91310	57
4	73689	92326	74831	91823	75931	91301	56
5	73708	92318	74850	91815	75949	91292	55
6	73727	92310	74868	91806	75967	91283	54
7	73747	92302	74887	91798	75985	91274	53
8	73766	92293	74906	91789	76003	91266	52
9	73785	92285	74924	91781	76021	91257	51
10	9.73805	9.92277	9.74943	9.91772	9.76039	9.91248	50
11	73824	92269	74961	91763	76057	91239	49
12	73843	92260	74980	91755	76075	91230	48
13	73863	92252	74999	91746	76093	91221	47
14	73882	92244	75017	91738	76111	91212	46
15	73901	92235	75036	91729	76129	91203	45
16	73921	92227	75054	91720	76146	91194	44
17	73940	92219	75073	91712	76164	91185	43
18	73959	92211	75091	91703	76182	91176	42
19	73978	92202	75110	91695	76200	91167	41
20	9.73997	9.92194	9.75128	9.91686	9.76218	9.91158	40
21	74017	92186	75147	91677	76236	91149	39
22	74036	92177	75165	91669	76253	91141	38
23	74055	92169	75184	91660	76271	91132	37
24	74074	92161	75202	91651	76289	91123	36
25	74093	92152	75221	91643	76307	91114	35
26	74113	92144	75239	91634	76324	91105	34
27	74132	92136	75258	91625	76342	91096	33
28	74151	92127	75276	91617	76360	91087	32
29	74170	92119	75294	91608	76378	91078	31
30	9.74189	9.92111	9.75313	9.91599	9.76395	9.91069	30
31	74208	92102	75331	91591	76413	91060	29
32	74227	92094	75350	91582	76431	91051	28
33	74246	92086	75368	91573	76448	91042	27
34	74265	92077	75386	91565	76466	91033	26
35	74284	92069	75405	91556	76484	91023	25
36	74303	92060	75423	91547	76501	91014	24
37	74322	92052	75441	91538	76519	91005	23
38	74341	92044	75459	91530	76537	90996	22
39	74360	92035	75478	91521	76554	90987	21
40	9.74379	9.92027	9.75496	9.91512	9.76572	9.90978	20
41	74398	92018	75514	91504	76590	90969	19
42	74417	92010	75533	91495	76607	90960	18
43	74436	92002	75551	91486	76625	90951	17
44	74455	91993	75569	91477	76642	90942	16
45	74474	91985	75587	91469	76660	90933	15
46	74493	91976	75605	91460	76677	90924	14
47	74512	91968	75624	91451	76695	90915	13
48	74531	91959	75642	91442	76712	90906	12
49	74549	91951	75660	91433	76730	90896	11
50	9.74568	9.91942	9.75678	9.91425	9.76747	9.90887	10
51	74587	91934	75696	91416	76765	90878	9
52	74606	91925	75714	91407	76782	90869	8
53	74625	91917	75733	91398	76800	90860	7
54	74644	91908	75751	91389	76817	90851	6
55	74662	91900	75769	91381	76835	90842	5
56	74681	91891	75787	91372	76852	90832	4
57	74700	91883	75805	91363	76870	90823	3
58	74719	91874	75823	91354	76887	90814	2
59	74737	91866	75841	91345	76904	90805	1
60	74756	91857	75859	91336	76922	90796	0
′	Cosine	Sine	Cosine	Sine	Cosine	Sine	′
	56°		55°		54°		

TABLE 2. LOGARITHMIC SINES AND COSINES

′	36° Sine	36° Cosine	37° Sine	37° Cosine	38° Sine	38° Cosine	′
0	9.76922	9.90796	9.77946	9.90235	9.78934	9.89653	60
1	76939	90787	77963	90225	78950	89643	59
2	76957	90777	77980	90216	78967	89633	58
3	76974	90768	77997	90206	78983	89624	57
4	76991	90759	78013	90197	78999	89614	56
5	77009	90750	78030	90187	79015	89604	55
6	77026	90741	78047	90178	79031	89594	54
7	77043	90731	78063	90168	79047	89584	53
8	77061	90722	78080	90159	79063	89574	52
9	77078	90713	78097	90149	79079	89564	51
10	9.77095	9.90704	9.78113	9.90139	9.79095	9.89554	50
11	77112	90694	78130	90130	79111	89544	49
12	77130	90685	78147	90120	79128	89534	48
13	77147	90676	78163	90111	79144	89524	47
14	77164	90667	78180	90101	79160	89514	46
15	77181	90657	78197	90091	79176	89504	45
16	77199	90648	78213	90082	79192	89495	44
17	77216	90639	78230	90072	79208	89485	43
18	77233	90630	78246	90063	79224	89475	42
19	77250	90620	78263	90053	79240	89465	41
20	9.77268	9.90611	9.78280	9.90043	9.79256	9.89455	40
21	77285	90602	78296	90034	79272	89445	39
22	77302	90592	78313	90024	79288	89435	38
23	77319	90583	78329	90014	79304	89425	37
24	77336	90574	78346	90005	79319	89415	36
25	77353	90565	78362	89995	79335	89405	35
26	77370	90555	78379	89985	79351	89395	34
27	77387	90546	78395	89976	79367	89385	33
28	77405	90537	78412	89966	79383	89375	32
29	77422	90527	78428	89956	79399	89364	31
30	9.77439	9.90518	9.78445	9.89947	9.79415	9.89354	30
31	77456	90509	78461	89937	79431	89344	29
32	77473	90499	78478	89927	79447	89334	28
33	77490	90490	78494	89918	79463	89324	27
34	77507	90480	78510	89908	79478	89314	26
35	77524	90471	78527	89898	79494	89304	25
36	77541	90462	78543	89888	79510	89294	24
37	77558	90452	78560	89879	79526	89284	23
38	77575	90443	78576	89869	79542	89274	22
39	77592	90434	78592	89859	79558	89264	21
40	9.77609	9.90424	9.78609	9.89849	9.79573	9.89254	20
41	77626	90415	78625	89840	79589	89244	19
42	77643	90405	78642	89830	79605	89233	18
43	77660	90396	78658	89820	79621	89223	17
44	77677	90386	78674	89810	79636	89213	16
45	77694	90377	78691	89801	79652	89203	15
46	77711	90368	78707	89791	79668	89193	14
47	77728	90358	78723	89781	79684	89183	13
48	77744	90349	78739	89771	79699	89173	12
49	77761	90339	78756	89761	79715	89162	11
50	9.77778	9.90330	9.78772	9.89752	9.79731	9.89152	10
51	77795	90320	78788	89742	79746	89142	9
52	77812	90311	78805	89732	79762	89132	8
53	77829	90301	78821	89722	79778	89122	7
54	77846	90292	78837	89712	79793	89112	6
55	77862	90282	78853	89702	79809	89101	5
56	77879	90273	78869	89693	79825	89091	4
57	77896	90263	78886	89683	79840	89081	3
58	77913	90254	78902	89673	79856	89071	2
59	77930	90244	78918	89663	79872	89060	1
60	77946	90235	78934	89653	79887	89050	0
′	Cosine	Sine	Cosine	Sine	Cosine	Sine	′
	53°		52°		51°		

TABLE 2. LOGARITHMIC SINES AND COSINES

′	39°		40°		41°		′
	Sine	Cosine	Sine	Cosine	Sine	Cosine	
0	9.79887	9.89050	9.80807	9.88425	9.81694	9.87778	60
1	79903	89040	80822	88415	81709	87767	59
2	79918	89030	80837	88404	81723	87756	58
3	79934	89020	80852	88394	81738	87745	57
4	79950	89009	80867	88383	81752	87734	56
5	79965	88999	80882	88372	81767	87723	55
6	79981	88989	80897	88362	81781	87712	54
7	79996	88978	80912	88351	81796	87701	53
8	80012	88968	80927	88340	81810	87690	52
9	80027	88958	80942	88330	81825	87679	51
10	9.80043	9.88948	9.80957	9.88319	9.81839	9.87668	50
11	80058	88937	80972	88308	81854	87657	49
12	80074	88927	80987	88298	81868	87646	48
13	80089	88917	81002	88287	81882	87635	47
14	80105	88906	81017	88276	81897	87624	46
15	80120	88896	81032	88266	81911	87613	45
16	80136	88886	81047	88255	81926	87601	44
17	80151	88875	81061	88244	81940	87590	43
18	80166	88865	81076	88234	81955	87579	42
19	80182	88855	81091	88223	81969	87568	41
20	9.80197	9.88844	9.81106	9.88212	9.81983	9.87557	40
21	80213	88834	81121	88201	81998	87546	39
22	80228	88824	81136	88191	82012	87535	38
23	80244	88813	81151	88180	82026	87524	37
24	80259	88803	81166	88169	82041	87513	36
25	80274	88793	81180	88158	82055	87501	35
26	80290	88782	81195	88148	82069	87490	34
27	80305	88772	81210	88137	82084	87479	33
28	80320	88761	81225	88126	82098	87468	32
29	80336	88751	81240	88115	82112	87457	31
30	9.80351	9.88741	9.81254	9.88105	9.82126	9.87446	30
31	80366	88730	81269	88094	82141	87434	29
32	80382	88720	81284	88083	82155	87423	28
33	80397	88709	81299	88072	82169	87412	27
34	80412	88699	81314	88061	82184	87401	26
35	80428	88688	81328	88051	82198	87390	25
36	80443	88678	81343	88040	82212	87378	24
37	80458	88668	81358	88029	82226	87367	23
38	80473	88657	81372	88018	82240	87356	22
39	80489	88647	81387	88007	82255	87345	21
40	9.80504	9.88636	9.81402	9.87996	9.82269	9.87334	20
41	80519	88626	81417	87985	82283	87322	19
42	80534	88615	81431	87975	82297	87311	18
43	80550	88605	81446	87964	82311	87300	17
44	80565	88594	81461	87953	82326	87288	16
45	80580	88584	81475	87942	82340	87277	15
46	80595	88573	81490	87931	82354	87266	14
47	80610	88563	81505	87920	82368	87255	13
48	80625	88552	81519	87909	82382	87243	12
49	80641	88542	81534	87898	82396	87232	11
50	9.80656	9.88531	9.81549	9.87887	9.82410	9.87221	10
51	80671	88521	81563	87877	82424	87209	9
52	80686	88510	81578	87866	82439	87198	8
53	80701	88499	81592	87855	82453	87187	7
54	80716	88489	81607	87844	82467	87175	6
55	80731	88478	81622	87833	82481	87164	5
56	80746	88468	81636	87822	82495	87153	4
57	80762	88457	81651	87811	82509	87141	3
58	80777	88447	81665	87800	82523	87130	2
59	80792	88436	81680	87789	82537	87119	1
60	80807	88425	81694	87778	82551	87107	0
′	Cosine	Sine	Cosine	Sine	Cosine	Sine	′
	50°		49°		48°		

TABLE 2. LOGARITHMIC SINES AND COSINES

′	42°		43°		44°		′
	Sine	Cosine	Sine	Cosine	Sine	Cosine	
0	9.82551	9.87107	9.83378	9.86413	9.84177	9.85693	60
1	82565	87096	83392	86401	84190	85681	59
2	82579	87085	83405	86389	84203	85669	58
3	82593	87073	83419	86377	84216	85657	57
4	82607	87062	83432	86366	84229	85645	56
5	82621	87050	83446	86354	84242	85632	55
6	82635	87039	83459	86342	84255	85620	54
7	82649	87028	83473	86330	84269	85608	53
8	82663	87016	83486	86318	84282	85596	52
9	82677	87005	83500	86306	84295	85583	51
10	9.82691	9.86993	9.83513	9.86295	9.84308	9.85571	50
11	82705	86982	83527	86283	84321	85559	49
12	82719	86970	83540	86271	84334	85547	48
13	82733	86959	83554	86259	84347	85534	47
14	82747	86947	83567	86247	84360	85522	46
15	82761	86936	83581	86235	84373	85510	45
16	82775	86924	83594	86223	84385	85497	44
17	82788	86913	83608	86211	84398	85485	43
18	82802	86902	83621	86200	84411	85473	42
19	82816	86890	83634	86188	84424	85460	41
20	9.82830	9.86879	9.83648	9.86176	9.84437	9.85448	40
21	82844	86867	83661	86164	84450	85436	39
22	82858	86855	83674	86152	84463	85423	38
23	82872	86844	83688	86140	84476	85411	37
24	82885	86832	83701	86128	84489	85399	36
25	82899	86821	83715	86116	84502	85386	35
26	82913	86809	83728	86104	84515	85374	34
27	82927	86798	83741	86092	84528	85361	33
28	82941	86786	83755	86080	84540	85349	32
29	82955	86775	83768	86068	84553	85337	31
30	9.82968	9.86763	9.83781	9.86056	9.84566	9.85324	30
31	82982	86752	83795	86044	84579	85312	29
32	82996	86740	83808	86032	84592	85299	28
33	83010	86728	83821	86020	84605	85287	27
34	83023	86717	83834	86008	84618	85274	26
35	83037	86705	83848	85996	84630	85262	25
36	83051	86694	83861	85984	84643	85250	24
37	83065	86682	83874	85972	84656	85237	23
38	83078	86670	83887	85960	84669	85225	22
39	83092	86659	83901	85948	84682	85212	21
40	9.83106	9.86647	9.83914	9.85936	9.84694	9.85200	20
41	83120	86635	83927	85924	84707	85187	19
42	83133	86624	83940	85912	84720	85175	18
43	83147	86612	83954	85900	84733	85162	17
44	83161	86600	83967	85888	84745	85150	16
45	83174	86589	83980	85876	84758	85137	15
46	83188	86577	83993	85864	84771	85125	14
47	83202	86565	84006	85851	84784	85112	13
48	83215	86554	84020	85839	84796	85100	12
49	83229	86542	84033	85827	84809	85087	11
50	9.83242	9.86530	9.84046	9.85815	9.84822	9.85074	10
51	83256	86518	84059	85803	84835	85062	9
52	83270	86507	84072	85791	84847	85049	8
53	83283	86495	84085	85779	84860	85037	7
54	83297	86483	84098	85766	84873	85024	6
55	83310	86472	84112	85754	84885	85012	5
56	83324	86460	84125	85742	84898	84999	4
57	83338	86448	84138	85730	84911	84986	3
58	83351	86436	84151	85718	84923	84974	2
59	83365	86425	84164	85706	84936	84961	1
60	83378	86413	84177	85693	84949	84949	0
′	Cosine	Sine	Cosine	Sine	Cosine	Sine	′
	47°		46°		45°		

TABLE 3. LOGARITHMIC TANGENTS AND COTANGENTS

′	0° Tan	0° Cotan	1° Tan	1° Cotan	2° Tan	2° Cotan	′
0	−∞	∞	8.24192	11.75808	8.54308	11.45692	60
1	6.46373	13.53627	24910	75090	54669	45331	59
2	76476	23524	25616	74384	55027	44973	58
3	94085	05915	26312	73688	55382	44618	57
4	7.06579	12.93421	26996	73004	55734	44266	56
5	16270	83730	27669	72331	56083	43917	55
6	24188	75812	28332	71668	56429	43571	54
7	30882	69118	28986	71014	56773	43227	53
8	36682	63318	29629	70371	57114	42886	52
9	41797	58203	30263	69737	57452	42548	51
10	7.46373	12.53627	8.30888	11.69112	8.57788	11.42212	50
11	50512	49488	31505	68495	58121	41879	49
12	54291	45709	32112	67888	58451	41549	48
13	57767	42233	32711	67289	58779	41221	47
14	60986	39014	33302	66698	59105	40895	46
15	63982	36018	33886	66114	59428	40572	45
16	66785	33215	34461	65539	59749	40251	44
17	69418	30582	35029	64971	60068	39932	43
18	71900	28100	35590	64410	60384	39616	42
19	74248	25752	36143	63857	60698	39302	41
20	7.76476	12.23524	8.36689	11.63311	8.61009	11.38991	40
21	78595	21405	37229	62771	61319	38681	39
22	80615	19385	37762	62238	61626	38374	38
23	82546	17454	38289	61711	61931	38069	37
24	84394	15606	38809	61191	62234	37766	36
25	86167	13833	39323	60677	62535	37465	35
26	87871	12129	39832	60168	62834	37166	34
27	89510	10490	40334	59666	63131	36869	33
28	91089	08911	40830	59170	63426	36574	32
29	92613	07387	41321	58679	63718	36282	31
30	7.94086	12.05914	8.41807	11.58193	8.64009	11.35991	30
31	95510	04490	42287	57713	64298	35702	29
32	96889	03111	42762	57238	64585	35415	28
33	98225	01775	43232	56768	64870	35130	27
34	99522	00478	43696	56304	65154	34846	26
35	8.00781	11.99219	44156	55844	65435	34565	25
36	02004	97996	44611	55389	65715	34285	24
37	03194	96806	45061	54939	65993	34007	23
38	04353	95647	45507	54493	66269	33731	22
39	05481	94519	45948	54052	66543	33457	21
40	8.06581	11.93419	8.46385	11.53615	8.66816	11.33184	20
41	07653	92347	46817	53183	67087	32913	19
42	08700	91300	47245	52755	67356	32644	18
43	09722	90278	47669	52331	67624	32376	17
44	10720	89280	48089	51911	67890	32110	16
45	11696	88304	48505	51495	68154	31846	15
46	12651	87349	48917	51083	68417	31583	14
47	13585	86415	49325	50675	68678	31322	13
48	14500	85500	49729	50271	68938	31062	12
49	15395	84605	50130	49870	69196	30804	11
50	8.16273	11.83727	8.50527	11.49473	8.69453	11.30547	10
51	17133	82867	50920	49080	69708	30292	9
52	17976	82024	51310	48690	69962	30038	8
53	18804	81196	51696	48304	70214	29786	7
54	19616	80384	52079	47921	70465	29535	6
55	20413	79587	52459	47541	70714	29286	5
56	21195	78805	52835	47165	70962	29038	4
57	21964	78036	53208	46792	71208	28792	3
58	22720	77280	53578	46422	71453	28547	2
59	23462	76538	53945	46055	71697	28303	1
60	24192	75808	54308	45692	71940	28060	0
′	Cotan	Tan	Cotan	Tan	Cotan	Tan	′
	89°		88°		87°		

TABLE 3. LOGARITHMIC TANGENTS AND COTANGENTS

'	3°		4°		5°		'
	Tan	Cotan	Tan	Cotan	Tan	Cotan	
0	8.71940	11.28060	8.84464	11.15536	8.94195	11.05805	60
1	72181	27819	84646	15354	94340	05660	59
2	72420	27580	84826	15174	94485	05515	58
3	72659	27341	85006	14994	94630	05370	57
4	72896	27104	85185	14815	94773	05227	56
5	73132	26868	85363	14637	94917	05083	55
6	73366	26634	85540	14460	95060	04940	54
7	73600	26400	85717	14283	95202	04798	53
8	73832	26168	85893	14107	95344	04656	52
9	74063	25937	86069	13931	95486	04514	51
10	8.74292	11.25708	8.86243	11.13757	8.95627	11.04373	50
11	74521	25479	86417	13583	95767	04233	49
12	74748	25252	86591	13409	95908	04092	48
13	74974	25026	86763	13237	96047	03953	47
14	75199	24801	86935	13065	96187	03813	46
15	75423	24577	87106	12894	96325	03675	45
16	75645	24355	87277	12723	96464	03536	44
17	75867	24133	87447	12553	96602	03398	43
18	76087	23913	87616	12384	96739	03261	42
19	76306	23694	87785	12215	96877	03123	41
20	8.76525	11.23475	8.87953	11.12047	8.97013	11.02987	40
21	76742	23258	88120	11880	97150	02850	39
22	76958	23042	88287	11713	97285	02715	38
23	77173	22827	88453	11547	97421	02579	37
24	77387	22613	88618	11382	97556	02444	36
25	77600	22400	88783	11217	97691	02309	35
26	77811	22189	88948	11052	97825	02175	34
27	78022	21978	89111	10889	97959	02041	33
28	78232	21768	89274	10726	98092	01908	32
29	78441	21559	89437	10563	98225	01775	31
30	8.78649	11.21351	8.89598	11.10402	8.98358	11.01642	30
31	78855	21145	89760	10240	98490	01510	29
32	79061	20939	89920	10080	98622	01378	28
33	79266	20734	90080	09920	98753	01247	27
34	79470	20530	90240	09760	98884	01116	26
35	79673	20327	90399	09601	99015	00985	25
36	79875	20125	90557	09443	99145	00855	24
37	80076	19924	90715	09285	99275	00725	23
38	80277	19723	90872	09128	99405	00595	22
39	80476	19524	91029	08971	99534	00466	21
40	8.80674	11.19326	8.91185	11.08815	8.99662	11.00338	20
41	80872	19128	91340	08660	99791	00209	19
42	81068	18932	91495	08505	99919	00081	18
43	81264	18736	91650	08350	9.00046	10.99954	17
44	81459	18541	91803	08197	00174	99826	16
45	81653	18347	91957	08043	00301	99699	15
46	81846	18154	92110	07890	00427	99573	14
47	82038	17962	92262	07738	00553	99447	13
48	82230	17770	92414	07586	00679	99321	12
49	82420	17580	92565	07435	00805	99195	11
50	8.82610	11.17390	8.92716	11.07284	9.00930	10.99070	10
51	82799	17201	92866	07134	01055	98945	9
52	82987	17013	93016	06984	01179	98821	8
53	83175	16825	93165	06835	01303	98697	7
54	83361	16639	93313	06687	01427	98573	6
55	83547	16453	93462	06538	01550	98450	5
56	83732	16268	93609	06391	01673	98327	4
57	83916	16084	93756	06244	01796	98204	3
58	84100	15900	93903	06097	01918	98082	2
59	84282	15718	94049	05951	02040	97960	1
60	84464	15536	94195	05805	02162	97838	0
'	Cotan	Tan	Cotan	Tan	Cotan	Tan	'
	86°		85°		84°		

TABLE 3. LOGARITHMIC TANGENTS AND COTANGENTS

′	6°		7°		8°		′
	Tan	Cotan	Tan	Cotan	Tan	Cotan	
0	9.02162	10.97838	9.08914	10.91086	9.14780	10.85220	60
1	02283	97717	09019	90981	14872	85128	59
2	02404	97596	09123	90877	14963	85037	58
3	02525	97475	09227	90773	15054	84946	57
4	02645	97355	09330	90670	15145	84855	56
5	02766	97234	09434	90566	15236	84764	55
6	02885	97115	09537	90463	15327	84673	54
7	03005	96995	09640	90360	15417	84583	53
8	03124	96876	09742	90258	15508	84492	52
9	03242	96758	09845	90155	15598	84402	51
10	9.03361	10.96639	9.09947	10.90053	9.15688	10.84312	50
11	03479	96521	10049	89951	15777	84223	49
12	03597	96403	10150	89850	15867	84133	48
13	03714	96286	10252	89748	15956	84044	47
14	03832	96168	10353	89647	16046	83954	46
15	03948	96052	10454	89546	16135	83865	45
16	04065	95935	10555	89445	16224	83776	44
17	04181	95819	10656	89344	16312	83688	43
18	04297	95703	10756	89244	16401	83599	42
19	04413	95587	10856	89144	16489	83511	41
20	9.04528	10.95472	9.10956	10.89044	9.16577	10.83423	40
21	04643	95357	11056	88944	16665	83335	39
22	04758	95242	11155	88845	16753	83247	38
23	04873	95127	11254	88746	16841	83159	37
24	04987	95013	11353	88647	16928	83072	36
25	05101	94899	11452	88548	17016	82984	35
26	05214	94786	11551	88449	17103	82897	34
27	05328	94672	11649	88351	17190	82810	33
28	05441	94559	11747	88253	17277	82723	32
29	05553	94447	11845	88155	17363	82637	31
30	9.05666	10.94334	9.11943	10.88057	9.17450	10.82550	30
31	05778	94222	12040	87960	17536	82464	29
32	05890	94110	12138	87862	17622	82378	28
33	06002	93998	12235	87765	17708	82292	27
34	06113	93887	12332	87668	17794	82206	26
35	06224	93776	12428	87572	17880	82120	25
36	06335	93665	12525	87475	17965	82035	24
37	06445	93555	12621	87379	18051	81949	23
38	06556	93444	12717	87283	18136	81864	22
39	06666	93334	12813	87187	18221	81779	21
40	9.06775	10.93225	9.12909	10.87091	9.18306	10.81694	20
41	06885	93115	13004	86996	18391	81609	19
42	06994	93006	13099	86901	18475	81525	18
43	07103	92897	13194	86806	18560	81440	17
44	07211	92789	13289	86711	18644	81356	16
45	07320	92680	13384	86616	18728	81272	15
46	07428	92572	13478	86522	18812	81188	14
47	07536	92464	13573	86427	18896	81104	13
48	07643	92357	13667	86333	18979	81021	12
49	07751	92249	13761	86239	19063	80937	11
50	9.07858	10.92142	9.13854	10.86146	9.19146	10.80854	10
51	07964	92036	13948	86052	19229	80771	9
52	08071	91929	14041	85959	19312	80688	8
53	08177	91823	14134	85866	19395	80605	7
54	08283	91717	14227	85773	19478	80522	6
55	08389	91611	14320	85680	19561	80439	5
56	08495	91505	14412	85588	19643	80357	4
57	08600	91400	14504	85496	19725	80275	3
58	08705	91295	14597	85403	19807	80193	2
59	08810	91190	14688	85312	19889	80111	1
60	08914	91086	14780	85220	19971	80029	0
′	Cotan	Tan	Cotan	Tan	Cotan	Tan	′
	83°		82°		81°		

TABLE 3. LOGARITHMIC TANGENTS AND COTANGENTS

′	9°		10°		11°		′
	Tan	Cotan	Tan	Cotan	Tan	Cotan	
0	9.19971	10.80029	9.24632	10.75368	9.28865	10.71135	60
1	20053	79947	24706	75294	28933	71067	59
2	20134	79866	24779	75221	29000	71000	58
3	20216	79784	24853	75147	29067	70933	57
4	20297	79703	24926	75074	29134	70866	56
5	20378	79622	25000	75000	29201	70799	55
6	20459	79541	25073	74927	29268	70732	54
7	20540	79460	25146	74854	29335	70665	53
8	20621	79379	25219	74781	29402	70598	52
9	20701	79299	25292	74708	29468	70532	51
10	9.20782	10.79218	9.25365	10.74635	9.29535	10.70465	50
11	20862	79138	25437	74563	29601	70399	49
12	20942	79058	25510	74490	29668	70332	48
13	21022	78978	25582	74418	29734	70266	47
14	21102	78898	25655	74345	29800	70200	46
15	21182	78818	25727	74273	29866	70134	45
16	21261	78739	25799	74201	29932	70068	44
17	21341	78659	25871	74129	29998	70002	43
18	21420	78580	25943	74057	30064	69936	42
19	21499	78501	26015	73985	30130	69870	41
20	9.21578	10.78422	9.26086	10.73914	9.30195	10.69805	40
21	21657	78343	26158	73842	30261	69739	39
22	21736	78264	26229	73771	30326	69674	38
23	21814	78186	26301	73699	30391	69609	37
24	21893	78107	26372	73628	30457	69543	36
25	21971	78029	26443	73557	30522	69478	35
26	22049	77951	26514	73486	30587	69413	34
27	22127	77873	26585	73415	30652	69348	33
28	22205	77795	26655	73345	30717	69283	32
29	22283	77717	26726	73274	30782	69218	31
30	9.22361	10.77639	9.26797	10.73203	9.30846	10.69154	30
31	22438	77562	26867	73133	30911	69089	29
32	22516	77484	26937	73063	30975	69025	28
33	22593	77407	27008	72992	31040	68960	27
34	22670	77330	27078	72922	31104	68896	26
35	22747	77253	27148	72852	31168	68832	25
36	22824	77176	27218	72782	31233	68767	24
37	22901	77099	27288	72712	31297	68703	23
38	22977	77023	27357	72643	31361	68639	22
39	23054	76946	27427	72573	31425	68575	21
40	9.23130	10.76870	9.27496	10.72504	9.31489	10.68511	20
41	23206	76794	27566	72434	31552	68448	19
42	23283	76717	27635	72365	31616	68384	18
43	23359	76641	27704	72296	31679	68321	17
44	23435	76565	27773	72227	31743	68257	16
45	23510	76490	27842	72158	31806	68194	15
46	23586	76414	27911	72089	31870	68130	14
47	23661	76339	27980	72020	31933	68067	13
48	23737	76263	28049	71951	31996	68004	12
49	23812	76188	28117	71883	32059	67941	11
50	9.23887	10.76113	9.28186	10.71814	9.32122	10.67878	10
51	23962	76038	28254	71746	32185	67815	9
52	24037	75963	28323	71677	32248	67752	8
53	24112	75888	28391	71609	32311	67689	7
54	24186	75814	28459	71541	32373	67627	6
55	24261	75739	28527	71473	32436	67564	5
56	24335	75665	28595	71405	32498	67502	4
57	24410	75590	28662	71338	32561	67439	3
58	24484	75516	28730	71270	32623	67377	2
59	24558	75442	28798	71202	32685	67315	1
60	24632	75368	28865	71135	32747	67253	0
′	Cotan	Tan	Cotan	Tan	Cotan	Tan	′
	80°		79°		78°		

232

TABLE 3. LOGARITHMIC TANGENTS AND COTANGENTS

′	12° Tan	12° Cotan	13° Tan	13° Cotan	14° Tan	14° Cotan	′
0	9.32747	10.67253	9.36336	10.63664	9.39677	10.60323	60
1	32810	67190	36394	63606	39731	60269	59
2	32872	67128	36452	63548	39785	60215	58
3	32933	67067	36509	63491	39838	60162	57
4	32995	67005	36566	63434	39892	60108	56
5	33057	66943	36624	63376	39945	60055	55
6	33119	66881	36681	63319	39999	60001	54
7	33180	66820	36738	63262	40052	59948	53
8	33242	66758	36795	63205	40106	59894	52
9	33303	66697	36852	63148	40159	59841	51
10	9.33365	10.66635	9.36909	10.63091	9.40212	10.59788	50
11	33426	66574	36966	63034	40266	59734	49
12	33487	66513	37023	62977	40319	59681	48
13	33548	66452	37080	62920	40372	59628	47
14	33609	66391	37137	62863	40425	59575	46
15	33670	66330	37193	62807	40478	59522	45
16	33731	66269	37250	62750	40531	59469	44
17	33792	66208	37306	62694	40584	59416	43
18	33853	66147	37363	62637	40636	59364	42
19	33913	66087	37419	62581	40689	59311	41
20	9.33974	10.66026	9.37476	10.62524	9.40742	10.59258	40
21	34034	65966	37532	62468	40795	59205	39
22	34095	65905	37588	62412	40847	59153	38
23	34155	65845	37644	62356	40900	59100	37
24	34215	65785	37700	62300	40952	59048	36
25	34276	65724	37756	62244	41005	58995	35
26	34336	65664	37812	62188	41057	58943	34
27	34396	65604	37868	62132	41109	58891	33
28	34456	65544	37924	62076	41161	58839	32
29	34516	65484	37980	62020	41214	58786	31
30	9.34576	10.65424	9.38035	10.61965	9.41266	10.58734	30
31	34635	65365	38091	61909	41318	58682	29
32	34695	65305	38147	61853	41370	58630	28
33	34755	65245	38202	61798	41422	58578	27
34	34814	65186	38257	61743	41474	58526	26
35	34874	65126	38313	61687	41526	58474	25
36	34933	65067	38368	61632	41578	58422	24
37	34992	65008	38423	61577	41629	58371	23
38	35051	64949	38479	61521	41681	58319	22
39	35111	64889	38534	61466	41733	58267	21
40	9.35170	10.64830	9.38589	10.61411	9.41784	10.58216	20
41	35229	64771	38644	61356	41836	58164	19
42	35288	64712	38699	61301	41887	58113	18
43	35347	64653	38754	61246	41939	58061	17
44	35405	64595	38808	61192	41990	58010	16
45	35464	64536	38863	61137	42041	57959	15
46	35523	64477	38918	61082	42093	57907	14
47	35581	64419	38972	61028	42144	57856	13
48	35640	64360	39027	60973	42195	57805	12
49	35698	64302	39082	60918	42246	57754	11
50	9.35757	10.64243	9.39136	10.60864	9.42297	10.57703	10
51	35815	64185	39190	60810	42348	57652	9
52	35873	64127	39245	60755	42399	57601	8
53	35931	64069	39299	60701	42450	57550	7
54	35989	64011	39353	60647	42501	57499	6
55	36047	63953	39407	60593	42552	57448	5
56	36105	63895	39461	60539	42603	57397	4
57	36163	63837	39515	60485	42653	57347	3
58	36221	63779	39569	60431	42704	57296	2
59	36279	63721	39623	60377	42755	57245	1
60	36336	63664	39677	60323	42805	57195	0

′	Cotan	Tan	Cotan	Tan	Cotan	Tan	′
	77°		76°		75°		

233

TABLE 3. LOGARITHMIC TANGENTS AND COTANGENTS

′	15°		16°		17°		′
	Tan	Cotan	Tan	Cotan	Tan	Cotan	
0	9.42805	10.57195	9.45750	10.54250	9.48534	10.51466	60
1	42856	57144	45797	54203	48579	51421	59
2	42906	57094	45845	54155	48624	51376	58
3	42957	57043	45892	54108	48669	51331	57
4	43007	56993	45940	54060	48714	51286	56
5	43057	56943	45987	54013	48759	51241	55
6	43108	56892	46035	53965	48804	51196	54
7	43158	56842	46082	53918	48849	51151	53
8	43208	56792	46130	53870	48894	51106	52
9	43258	56742	46177	53823	48939	51061	51
10	9.43308	10.56692	9.46224	10.53776	9.48984	10.51016	50
11	43358	56642	46271	53729	49029	50971	49
12	43408	56592	46319	53681	49073	50927	48
13	43458	56542	46366	53634	49118	50882	47
14	43508	56492	46413	53587	49163	50837	46
15	43558	56442	46460	53540	49207	50793	45
16	43607	56393	46507	53493	49252	50748	44
17	43657	56343	46554	53446	49296	50704	43
18	43707	56293	46601	53399	49341	50659	42
19	43756	56244	46648	53352	49385	50615	41
20	9.43806	10.56194	9.46694	10.53306	9.49430	10.50570	40
21	43855	56145	46741	53259	49474	50526	39
22	43905	56095	46788	53212	49519	50481	38
23	43954	56046	46835	53165	49563	50437	37
24	44004	55996	46881	53119	49607	50393	36
25	44053	55947	46928	53072	49652	50348	35
26	44102	55898	46975	53025	49696	50304	34
27	44151	55849	47021	52979	49740	50260	33
28	44201	55799	47068	52932	49784	50216	32
29	44250	55750	47114	52886	49828	50172	31
30	9.44299	10.55701	9.47160	10.52840	9.49872	10.50128	30
31	44348	55652	47207	52793	49916	50084	29
32	44397	55603	47253	52747	49960	50040	28
33	44446	55554	47299	52701	50004	49996	27
34	44495	55505	47346	52654	50048	49952	26
35	44544	55456	47392	52608	50092	49908	25
36	44592	55408	47438	52562	50136	49864	24
37	44641	55359	47484	52516	50180	49820	23
38	44690	55310	47530	52470	50223	49777	22
39	44738	55262	47576	52424	50267	49733	21
40	9.44787	10.55213	9.47622	10.52378	9.50311	10.49689	20
41	44836	55164	47668	52332	50355	49645	19
42	44884	55116	47714	52286	50398	49602	18
43	44933	55067	47760	52240	50442	49558	17
44	44981	55019	47806	52194	50485	49515	16
45	45029	54971	47852	52148	50529	49471	15
46	45078	54922	47897	52103	50572	49428	14
47	45126	54874	47943	52057	50616	49384	13
48	45174	54826	47989	52011	50659	49341	12
49	45222	54778	48035	51965	50703	49297	11
50	9.45271	10.54729	9.48080	10.51920	9.50746	10.49254	10
51	45319	54681	48126	51874	50789	49211	9
52	45367	54633	48171	51829	50833	49167	8
53	45415	54585	48217	51783	50876	49124	7
54	45463	54537	48262	51738	50919	49081	6
55	45511	54489	48307	51693	50962	49038	5
56	45559	54441	48353	51647	51005	48995	4
57	45606	54394	48398	51602	51048	48952	3
58	45654	54346	48443	51557	51092	48908	2
59	45702	54298	48489	51511	51135	48865	1
60	45750	54250	48534	51466	51178	48822	0
′	Cotan	Tan	Cotan	Tan	Cotan	Tan	′
	74°		73°		72°		

234

TABLE 3. LOGARITHMIC TANGENTS AND COTANGENTS

′	18°		19°		20°		′
	Tan	Cotan	Tan	Cotan	Tan	Cotan	
0	9.51178	10.48822	9.53697	10.46303	9.56107	10.43893	60
1	51221	48779	53738	46262	56146	43854	59
2	51264	48736	53779	46221	56185	43815	58
3	51306	48694	53820	46180	56224	43776	57
4	51349	48651	53861	46139	56264	43736	56
5	51392	48608	53902	46098	56303	43697	55
6	51435	48565	53943	46057	56342	43658	54
7	51478	48522	53984	46016	56381	43619	53
8	51520	48480	54025	45975	56420	43580	52
9	51563	48437	54065	45935	56459	43541	51
10	9.51606	10.48394	9.54106	10.45894	9.56498	10.43502	50
11	51648	48352	54147	45853	56537	43463	49
12	51691	48309	54187	45813	56576	43424	48
13	51734	48266	54228	45772	56615	43385	47
14	51776	48224	54269	45731	56654	43346	46
15	51819	48181	54309	45691	56693	43307	45
16	51861	48139	54350	45650	56732	43268	44
17	51903	48097	54390	45610	56771	43229	43
18	51946	48054	54431	45569	56810	43190	42
19	51988	48012	54471	45529	56849	43151	41
20	9.52031	10.47969	9.54512	10.45488	9.56887	10.43113	40
21	52073	47927	54552	45448	56926	43074	39
22	52115	47885	54593	45407	56965	43035	38
23	52157	47843	54633	45367	57004	42996	37
24	52200	47800	54673	45327	57042	42958	36
25	52242	47758	54714	45286	57081	42919	35
26	52284	47716	54754	45246	57120	42880	34
27	52326	47674	54794	45206	57158	42842	33
28	52368	47632	54835	45165	57197	42803	32
29	52410	47590	54875	45125	57235	42765	31
30	9.52452	10.47548	9.54915	10.45085	9.57274	10.42726	30
31	52494	47506	54955	45045	57312	42688	29
32	52536	47464	54995	45005	57351	42649	28
33	52578	47422	55035	44965	57389	42611	27
34	52620	47380	55075	44925	57428	42572	26
35	52661	47339	55115	44885	57466	42534	25
36	52703	47297	55155	44845	57504	42496	24
37	52745	47255	55195	44805	57543	42457	23
38	52787	47213	55235	44765	57581	42419	22
39	52829	47171	55275	44725	57619	42381	21
40	9.52870	10.47130	9.55315	10.44685	9.57658	10.42342	20
41	52912	47088	55355	44645	57696	42304	19
42	52953	47047	55395	44605	57734	42266	18
43	52995	47005	55434	44566	57772	42228	17
44	53037	46963	55474	44526	57810	42190	16
45	53078	46922	55514	44486	57849	42151	15
46	53120	46880	55554	44446	57887	42113	14
47	53161	46839	55593	44407	57925	42075	13
48	53202	46798	55633	44367	57963	42037	12
49	53244	46756	55673	44327	58001	41999	11
50	9.53285	10.46715	9.55712	10.44288	9.58039	10.41961	10
51	53327	46673	55752	44248	58077	41923	9
52	53368	46632	55791	44209	58115	41885	8
53	53409	46591	55831	44169	58153	41847	7
54	53450	46550	55870	44130	58191	41809	6
55	53492	46508	55910	44090	58229	41771	5
56	53533	46467	55949	44051	58267	41733	4
57	53574	46426	55989	44011	58304	41696	3
58	53615	46385	56028	43972	58342	41658	2
59	53656	46344	56067	43933	58380	41620	1
60	53697	46303	56107	43893	58418	41582	0
′	Cotan	Tan	Cotan	Tan	Cotan	Tan	′
	71°		70°		69°		

TABLE 3. LOGARITHMIC TANGENTS AND COTANGENTS

′	21°		22°		23°		′
	Tan	Cotan	Tan	Cotan	Tan	Cotan	
0	9.58418	10.41582	9.60641	10.39359	9.62785	10.37215	60
1	58455	41545	60677	39323	62820	37180	59
2	58493	41507	60714	39286	62855	37145	58
3	58531	41469	60750	39250	62890	37110	57
4	58569	41431	60786	39214	62926	37074	56
5	58606	41394	60823	39177	62961	37039	55
6	58644	41356	60859	39141	62996	37004	54
7	58681	41319	60895	39105	63031	36969	53
8	58719	41281	60931	39069	63066	36934	52
9	58757	41243	60967	39033	63101	36899	51
10	9.58794	10.41206	9.61004	10.38996	9.63135	10.36865	50
11	58832	41168	61040	38960	63170	36830	49
12	58869	41131	61076	38924	63205	36795	48
13	58907	41093	61112	38888	63240	36760	47
14	58944	41056	61148	38852	63275	36725	46
15	58981	41019	61184	38816	63310	36690	45
16	59019	40981	61220	38780	63345	36655	44
17	59056	40944	61256	38744	63379	36621	43
18	59094	40906	61292	38708	63414	36586	42
19	59131	40869	61328	38672	63449	36551	41
20	9.59168	10.40832	9.61364	10.38636	9.63484	10.36516	40
21	59205	40795	61400	38600	63519	36481	39
22	59243	40757	61436	38564	63553	36447	38
23	59280	40720	61472	38528	63588	36412	37
24	59317	40683	61508	38492	63623	36377	36
25	59354	40646	61544	38456	63657	36343	35
26	59391	40609	61579	38421	63692	36308	34
27	59429	40571	61615	38385	63726	36274	33
28	59466	40534	61651	38349	63761	36239	32
29	59503	40497	61687	38313	63796	36204	31
30	9.59540	10.40460	9.61722	10.38278	9.63830	10.36170	30
31	59577	40423	61758	38242	63865	36135	29
32	59614	40386	61794	38206	63899	36101	28
33	59651	40349	61830	38170	63934	36066	27
34	59688	40312	61865	38135	63968	36032	26
35	59725	40275	61901	38099	64003	35997	25
36	59762	40238	61936	38064	64037	35963	24
37	59799	40201	61972	38028	64072	35928	23
38	59835	40165	62008	37992	64106	35894	22
39	59872	40128	62043	37957	64140	35860	21
40	9.59909	10.40091	9.62079	10.37921	9.64175	10.35825	20
41	59946	40054	62114	37886	64209	35791	19
42	59983	40017	62150	37850	64243	35757	18
43	60019	39981	62185	37815	64278	35722	17
44	60056	39944	62221	37779	64312	35688	16
45	60093	39907	62256	37744	64346	35654	15
46	60130	39870	62292	37708	64381	35619	14
47	60166	39834	62327	37673	64415	35585	13
48	60203	39797	62362	37638	64449	35551	12
49	60240	39760	62398	37602	64483	35517	11
50	9.60276	10.39724	9.62433	10.37567	9.64517	10.35483	10
51	60313	39687	62468	37532	64552	35448	9
52	60349	39651	62504	37496	64586	35414	8
53	60386	39614	62539	37461	64620	35380	7
54	60422	39578	62574	37426	64654	35346	6
55	60459	39541	62609	37391	64688	35312	5
56	60495	39505	62645	37355	64722	35278	4
57	60532	39468	62680	37320	64756	35244	3
58	60568	39432	62715	37285	64790	35210	2
59	60605	39395	62750	37250	64824	35176	1
60	60641	39359	62785	37215	64858	35142	0
′	Cotan	Tan	Cotan	Tan	Cotan	Tan	′
	68°		67°		66°		

236

TABLE 3. LOGARITHMIC TANGENTS AND COTANGENTS

′	24° Tan	24° Cotan	25° Tan	25° Cotan	26° Tan	26° Cotan	′
0	9.64858	10.35142	9.66867	10.33133	9.68818	10.31182	60
1	64892	35108	66900	33100	68850	31150	59
2	64926	35074	66933	33067	68882	31118	58
3	64960	35040	66966	33034	68914	31086	57
4	64994	35006	66999	33001	68946	31054	56
5	65028	34972	67032	32968	68978	31022	55
6	65062	34938	67065	32935	69010	30990	54
7	65096	34904	67098	32902	69042	30958	53
8	65130	34870	67131	32869	69074	30926	52
9	65164	34836	67163	32837	69106	30894	51
10	9.65197	10.34803	9.67196	10.32804	9.69138	10.30862	50
11	65231	34769	67229	32771	69170	30830	49
12	65265	34735	67262	32738	69202	30798	48
13	65299	34701	67295	32705	69234	30766	47
14	65333	34667	67327	32673	69266	30734	46
15	65366	34634	67360	32640	69298	30702	45
16	65400	34600	67393	32607	69329	30671	44
17	65434	34566	67426	32574	69361	30639	43
18	65467	34533	67458	32542	69393	30607	42
19	65501	34499	67491	32509	69425	30575	41
20	9.65535	10.34465	9.67524	10.32476	9.69457	10.30543	40
21	65568	34432	67556	32444	69488	30512	39
22	65602	34398	67589	32411	69520	30480	38
23	65636	34364	67622	32378	69552	30448	37
24	65669	34331	67654	32346	69584	30416	36
25	65703	34297	67687	32313	69615	30385	35
26	65736	34264	67719	32281	69647	30353	34
27	65770	34230	67752	32248	69679	30321	33
28	65803	34197	67785	32215	69710	30290	32
29	65837	34163	67817	32183	69742	30258	31
30	9.65870	10.34130	9.67850	10.32150	9.69774	10.30226	30
31	65904	34096	67882	32118	69805	30195	29
32	65937	34063	67915	32085	69837	30163	28
33	65971	34029	67947	32053	69868	30132	27
34	66004	33996	67980	32020	69900	30100	26
35	66038	33962	68012	31988	69932	30068	25
36	66071	33929	68044	31956	69963	30037	24
37	66104	33896	68077	31923	69995	30005	23
38	66138	33862	68109	31891	70026	29974	22
39	66171	33829	68142	31858	70058	29942	21
40	9.66204	10.33796	9.68174	10.31826	9.70089	10.29911	20
41	66238	33762	68206	31794	70121	29879	19
42	66271	33729	68239	31761	70152	29848	18
43	66304	33696	68271	31729	70184	29816	17
44	66337	33663	68303	31697	70215	29785	16
45	66371	33629	68336	31664	70247	29753	15
46	66404	33596	68368	31632	70278	29722	14
47	66437	33563	68400	31600	70309	29691	13
48	66470	33530	68432	31568	70341	29659	12
49	66503	33497	68465	31535	70372	29628	11
50	9.66537	10.33463	9.68497	10.31503	9.70404	10.29596	10
51	66570	33430	68529	31471	70435	29565	9
52	66603	33397	68561	31439	70466	29534	8
53	66636	33364	68593	31407	70498	29502	7
54	66669	33331	68626	31374	70529	29471	6
55	66702	33298	68658	31342	70560	29440	5
56	66735	33265	68690	31310	70592	29408	4
57	66768	33232	68722	31278	70623	29377	3
58	66801	33199	68754	31246	70654	29346	2
59	66834	33166	68786	31214	70685	29315	1
60	66867	33133	68818	31182	70717	29283	0
′	Cotan	Tan	Cotan	Tan	Cotan	Tan	′
	65°		64°		63°		

TABLE 3. LOGARITHMIC TANGENTS AND COTANGENTS

′	27°		28°		29°		′
	Tan	Cotan	Tan	Cotan	Tan	Cotan	
0	9.70717	10.29283	9.72567	10.27433	9.74375	10.25625	60
1	70748	29252	72598	27402	74405	25595	59
2	70779	29221	72628	27372	74435	25565	58
3	70810	29190	72659	27341	74465	25535	57
4	70841	29159	72689	27311	74494	25506	66
5	70873	29127	72720	27280	74524	25476	55
6	70904	29096	72750	27250	74554	25446	54
7	70935	29065	72780	27220	74583	25417	53
8	70966	29034	72811	27189	74613	25387	52
9	70997	29003	72841	27159	74643	25357	51
10	9.71028	10.28972	9.72872	10.27128	9.74673	10.25327	50
11	71059	28941	72902	27098	74702	25298	49
12	71090	28910	72932	27068	74732	25268	48
13	71121	28879	72963	27037	74762	25238	47
14	71153	28847	72993	27007	74791	25209	46
15	71184	28816	73023	26977	74821	25179	45
16	71215	28785	73054	26946	74851	25149	44
17	71246	28754	73084	26916	74880	25120	43
18	71277	28723	73114	26886	74910	25090	42
19	71308	28692	73144	26856	74939	25061	41
20	9.71339	10.28661	9.73175	10.26825	9.74969	10.25031	40
21	71370	28630	73205	26795	74998	25002	39
22	71401	28599	73235	26765	75028	24972	38
23	71431	28569	73265	26735	75058	24942	37
24	71462	28538	73295	26705	75087	24913	36
25	71493	28507	73326	26674	75117	24883	35
26	71524	28476	73356	26644	75146	24854	34
27	71555	28445	73386	26614	75176	24824	33
28	71586	28414	73416	26584	75205	24795	32
29	71617	28383	73446	26554	75235	24765	31
30	9.71648	10.28352	9.73476	10.26524	9.75264	10.24736	30
31	71679	28321	73507	26493	75294	24706	29
32	71709	28291	73537	26463	75323	24677	28
33	71740	28260	73567	26433	75353	24647	27
34	71771	28229	73597	26403	75382	24618	26
35	71802	28198	73627	26373	75411	24589	25
36	71833	28167	73657	26343	75441	24559	24
37	71863	28137	73687	26313	75470	24530	23
38	71894	28106	73717	26283	75500	24500	22
39	71925	28075	73747	26253	75529	24471	21
40	9.71955	10.28045	9.73777	10.26223	9.75558	10.24442	20
41	71986	28014	73807	26193	75588	24412	19
42	72017	27983	73837	26163	75617	24383	18
43	72048	27952	73867	26133	75647	24353	17
44	72078	27922	73897	26103	75676	24324	16
45	72109	27891	73927	26073	75705	24295	15
46	72140	27860	73957	26043	75735	24265	14
47	72170	27830	73987	26013	75764	24236	13
48	72201	27799	74017	25983	75793	24207	12
49	72231	27769	74047	25953	75822	24178	11
50	9.72262	10.27738	9.74077	10.25923	9.75852	10.24148	10
51	72293	27707	74107	25893	75881	24119	9
52	72323	27677	74137	25863	75910	24090	8
53	72354	27646	74166	25834	75939	24061	7
54	72384	27616	74196	25804	75969	24031	6
55	72415	27585	74226	25774	75998	24002	5
56	72445	27555	74256	25744	76027	23973	4
57	72476	27524	74286	25714	76056	23944	3
58	72506	27494	74316	25684	76086	23914	2
59	72537	27463	74345	25655	76115	23885	1
60	72567	27433	74375	25625	76144	23856	0
′	Cotan	Tan	Cotan	Tan	Cotan	Tan	′
	62°		61°		60°		

238

TABLE 3. LOGARITHMIC TANGENTS AND COTANGENTS

′	30°		31°		32°		′
	Tan	Cotan	Tan	Cotan	Tan	Cotan	
0	9.76144	10.23856	9.77877	10.22123	9.79579	10.20421	60
1	76173	23827	77906	22094	79607	20393	59
2	76202	23798	77935	22065	79635	20365	58
3	76231	23769	77963	22037	79663	20337	57
4	76261	23739	77992	22008	79691	20309	56
5	76290	23710	78020	21980	79719	20281	55
6	76319	23681	78049	21951	79747	20253	54
7	76348	23652	78077	21923	79776	20224	53
8	76377	23623	78106	21894	79804	20196	52
9	76406	23594	78135	21865	79832	20168	51
10	9.76435	10.23565	9.78163	10.21837	9.79860	10.20140	50
11	76464	23536	78192	21808	79888	20112	49
12	76493	23507	78220	21780	79916	20084	48
13	76522	23478	78249	21751	79944	20056	47
14	76551	23449	78277	21723	79972	20028	46
15	76580	23420	78306	21694	80000	20000	45
16	76609	23391	78334	21666	80028	19972	44
17	76639	23361	78363	21637	80056	19944	43
18	76668	23332	78391	21609	80084	19916	42
19	76697	23303	78419	21581	80112	19888	41
20	9.76725	10.23275	9.78448	10.21552	9.80140	10.19860	40
21	76754	23246	78476	21524	80168	19832	39
22	76783	23217	78505	21495	80195	19805	38
23	76812	23188	78533	21467	80223	19777	37
24	76841	23159	78562	21438	80251	19749	36
25	76870	23130	78590	21410	80279	19721	35
26	76899	23101	78618	21382	80307	19693	34
27	76928	23072	78647	21353	80335	19665	33
28	76957	23043	78675	21325	80363	19637	32
29	76986	23014	78704	21296	80391	19609	31
30	9.77015	10.22985	9.78732	10.21268	9.80419	10.19581	30
31	77044	22956	78760	21240	80447	19553	29
32	77073	22927	78789	21211	80474	19526	28
33	77101	22899	78817	21183	80502	19498	27
34	77130	22870	78845	21155	80530	19470	26
35	77159	22841	78874	21126	80558	19442	25
36	77188	22812	78902	21098	80586	19414	24
37	77217	22783	78930	21070	80614	19386	23
38	77246	22754	78959	21041	80642	19358	22
39	77274	22726	78987	21013	80669	19331	21
40	9.77303	10.22697	9.79015	10.20985	9.80697	10.19303	20
41	77332	22668	79043	20957	80725	19275	19
42	77361	22639	79072	20928	80753	19247	18
43	77390	22610	79100	20900	80781	19219	17
44	77418	22582	79128	20872	80808	19192	16
45	77447	22553	79156	20844	80836	19164	15
46	77476	22524	79185	20815	80864	19136	14
47	77505	22495	79213	20787	80892	19108	13
48	77533	22467	79241	20759	80919	19081	12
49	77562	22438	79269	20731	80947	19053	11
50	9.77591	10.22409	9.79297	10.20703	9.80975	10.19025	10
51	77619	22381	79326	20674	81003	18997	9
52	77648	22352	79354	20646	81030	18970	8
53	77677	22323	79382	20618	81058	18942	7
54	77706	22294	79410	20590	81086	18914	6
55	77734	22266	79438	20562	81113	18887	5
56	77763	22237	79466	20534	81141	18859	4
57	77791	22209	79495	20505	81169	18831	3
58	77820	22180	79523	20477	81196	18804	2
59	77849	22151	79551	20449	81224	18776	1
60	77877	22123	79579	20421	81252	18748	0
	Cotan	Tan	Cotan	Tan	Cotan	Tan	′
′	59°		58°		57°		

TABLE 3. LOGARITHMIC TANGENTS AND COTANGENTS

′	33°		34°		35°		′
	Tan	Cotan	Tan	Cotan	Tan	Cotan	
0	9.81252	10.18748	9.82899	10.17101	9.84523	10.15477	60
1	81279	18721	82926	17074	84550	15450	59
2	81307	18693	82953	17047	84576	15424	58
3	81335	18665	82980	17020	84603	15397	57
4	81362	18638	83008	16992	84630	15370	56
5	81390	18610	83035	16965	84657	15343	55
6	81418	18582	83062	16938	84684	15316	54
7	81445	18555	83089	16911	84711	15289	53
8	81473	18527	83117	16883	84738	15262	52
9	81500	18500	83144	16856	84764	15236	51
10	9.81528	10.18472	9.83171	10.16829	9.84791	10.15209	50
11	81556	18444	83198	16802	84818	15182	49
12	81583	18417	83225	16775	84845	15155	48
13	81611	18389	83252	16748	84872	15128	47
14	81638	18362	83280	16720	84899	15101	46
15	81666	18334	83307	16693	84925	15075	45
16	81693	18307	83334	16666	84952	15048	44
17	81721	18279	83361	16639	84979	15021	43
18	81748	18252	83388	16612	85006	14994	42
19	81776	18224	83415	16585	85033	14967	41
20	9.81803	10.18197	9.83442	10.16558	9.85059	10.14941	40
21	81831	18169	83470	16530	85086	14914	39
22	81858	18142	83497	16503	85113	14887	38
23	81886	18114	83524	16476	85140	14860	37
24	81913	18087	83551	16449	85166	14834	36
25	81941	18059	83578	16422	85193	14807	35
26	81968	18032	83605	16395	85220	14780	34
27	81996	18004	83632	16368	85247	14753	33
28	82023	17977	83659	16341	85273	14727	32
29	82051	17949	83686	16314	85300	14700	31
30	9.82078	10.17922	9.83713	10.16287	9.85327	10.14673	30
31	82106	17894	83740	16260	85354	14646	29
32	82133	17867	83768	16232	85380	14620	28
33	82161	17839	83795	16205	85407	14593	27
34	82188	17812	83822	16178	85434	14566	26
35	82215	17785	83849	16151	85460	14540	25
36	82243	17757	83876	16124	85487	14513	24
37	82270	17730	83903	16097	85514	14486	23
38	82298	17702	83930	16070	85540	14460	22
39	82325	17675	83957	16043	85567	14433	21
40	9.82352	10.17648	9.83984	10.16016	9.85594	10.14406	20
41	82380	17620	84011	15989	85620	14380	19
42	82407	17593	84038	15962	85647	14353	18
43	82435	17565	84065	15935	85674	14326	17
44	82462	17538	84092	15908	85700	14300	16
45	82489	17511	84119	15881	85727	14273	15
46	82517	17483	84146	15854	85754	14246	14
47	82544	17456	84173	15827	85780	14220	13
48	82571	17429	84200	15800	85807	14193	12
49	82599	17401	84227	15773	85834	14166	11
50	9.82626	10.17374	9.84254	10.15746	9.85860	10.14140	10
51	82653	17347	84280	15720	85887	14113	9
52	82681	17319	84307	15693	85913	14087	8
53	82708	17292	84334	15666	85940	14060	7
54	82735	17265	84361	15639	85967	14033	6
55	82762	17238	84388	15612	85993	14007	5
56	82790	17210	84415	15585	86020	13980	4
57	82817	17183	84442	15558	86046	13954	3
58	82844	17156	84469	15531	86073	13927	2
59	82871	17129	84496	15504	86100	13900	1
60	82899	17101	84523	15477	86126	13874	0
′	Cotan	Tan	Cotan	Tan	Cotan	Tan	′
	56°		55°		54°		

240

TABLE 3. LOGARITHMIC TANGENTS AND COTANGENTS

′	36°		37°		38°		′
	Tan	Cotan	Tan	Cotan	Tan	Cotan	
0	9.86126	10.13874	9.87711	10.12289	9.89281	10.10719	60
1	86153	13847	87738	12262	89307	10693	59
2	86179	13821	87764	12236	89333	10667	58
3	86206	13794	87790	12210	89359	10641	57
4	86232	13768	87817	12183	89385	10615	56
5	86259	13741	87843	12157	89411	10589	55
6	86285	13715	87869	12131	89437	10563	54
7	86312	13688	87895	12105	89463	10537	53
8	86338	13662	87922	12078	89489	10511	52
9	86365	13635	87948	12052	89515	10485	51
10	9.86392	10.13608	9.87974	10.12026	9.89541	10.10459	50
11	86418	13582	88000	12000	89567	10433	49
12	86445	13555	88027	11973	89593	10407	48
13	86471	13529	88053	11947	89619	10381	47
14	86498	13502	88079	11921	89645	10355	46
15	86524	13476	88105	11895	89671	10329	45
16	86551	13449	88131	11869	89697	10303	44
17	86577	13423	88158	11842	89723	10277	43
18	86603	13397	88184	11816	89749	10251	42
19	86630	13370	88210	11790	89775	10225	41
20	9.86656	10.13344	9.88236	10.11764	9.89801	10.10199	40
21	86683	13317	88262	11738	89827	10173	39
22	86709	13291	88289	11711	89853	10147	38
23	86736	13264	88315	11685	89879	10121	37
24	86762	13238	88341	11659	89905	10095	36
25	86789	13211	88367	11633	89931	10069	35
26	86815	13185	88393	11607	89957	10043	34
27	86842	13158	88420	11580	89983	10017	33
28	86868	13132	88446	11554	90009	09991	32
29	86894	13106	88472	11528	90035	09965	31
30	9.86921	10.13079	9.88498	10.11502	9.90061	10.09939	30
31	86947	13053	88524	11476	90086	09914	29
32	86974	13026	88550	11450	90112	09888	28
33	87000	13000	88577	11423	90138	09862	27
34	87027	12973	88603	11397	90164	09836	26
35	87053	12947	88629	11371	90190	09810	25
36	87079	12921	88655	11345	90216	09784	24
37	87106	12894	88681	11319	90242	09758	23
38	87132	12868	88707	11293	90268	09732	22
39	87158	12842	88733	11267	90294	09706	21
40	9.87185	10.12815	9.88759	10.11241	9.90320	10.09680	20
41	87211	12789	88786	11214	90346	09654	19
42	87238	12762	88812	11188	90371	09629	18
43	87264	12736	88838	11162	90397	09603	17
44	87290	12710	88864	11136	90423	09577	16
45	87317	12683	88890	11110	90449	09551	15
46	87343	12657	88916	11084	90475	09525	14
47	87369	12631	88942	11058	90501	09499	13
48	87396	12604	88968	11032	90527	09473	12
49	87422	12578	88994	11006	90553	09447	11
50	9.87448	10.12552	9.89020	10.10980	9.90578	10.09422	10
51	87475	12525	89046	10954	90604	09396	9
52	87501	12499	89073	10927	90630	09370	8
53	87527	12473	89099	10901	90656	09344	7
54	87554	12446	89125	10875	90682	09318	6
55	87580	12420	89151	10849	90708	09292	5
56	87606	12394	89177	10823	90734	09266	4
57	87633	12367	89203	10797	90759	09241	3
58	87659	12341	89229	10771	90785	09215	2
59	87685	12315	89255	10745	90811	09189	1
60	87711	12289	89281	10719	90837	09163	0
′	Cotan	Tan	Cotan	Tan	Cotan	Tan	′
	53°		52°		51°		

TABLE 3. LOGARITHMIC TANGENTS AND COTANGENTS

'	39°		40°		41°		'
	Tan	Cotan	Tan	Cotan	Tan	Cotan	
0	9.90837	10.09163	9.92381	10.07619	9.93916	10.06084	60
1	90863	09137	92407	07593	93942	06058	59
2	90889	09111	92433	07567	93967	06033	58
3	90914	09086	92458	07542	93993	06007	57
4	90940	09060	92484	07516	94018	05982	56
5	90966	09034	92510	07490	94044	05956	55
6	90992	09008	92535	07465	94069	05931	54
7	91018	08982	92561	07439	94095	05905	53
8	91043	08957	92587	07413	94120	05880	52
9	91069	08931	92612	07388	94146	05854	51
10	9.91095	10.08905	9.92638	10.07362	9.94171	10.05829	50
11	91121	08879	92663	07337	94197	05803	49
12	91147	08853	92689	07311	94222	05778	48
13	91172	08828	92715	07285	94248	05752	47
14	91198	08802	92740	07260	94273	05727	46
15	91224	08776	92766	07234	94299	05701	45
16	91250	08750	92792	07208	94324	05676	44
17	91276	08724	92817	07183	94350	05650	43
18	91301	08699	92843	07157	94375	05625	42
19	91327	08673	92868	07132	94401	05599	41
20	9.91353	10.08647	9.92894	10.07106	9.94426	10.05574	40
21	91379	08621	92920	07080	94452	05548	39
22	91404	08596	92945	07055	94477	05523	38
23	91430	08570	92971	07029	94503	05497	37
24	91456	08544	92996	07004	94528	05472	36
25	91482	08518	93022	06978	94554	05446	35
26	91507	08493	93048	06952	94579	05421	34
27	91533	08467	93073	06927	94604	05396	33
28	91559	08441	93099	06901	94630	05370	32
29	91585	08415	93124	06876	94655	05345	31
30	9.91610	10.08390	9.93150	10.06850	9.94681	10.05319	30
31	91636	08364	93175	06825	94706	05294	29
32	91662	08338	93201	06799	94732	05268	28
33	91688	08312	93227	06773	94757	05243	27
34	91713	08287	93252	06748	94783	05217	26
35	91739	08261	93278	06722	94808	05192	25
36	91765	08235	93303	06697	94834	05166	24
37	91791	08209	93329	06671	94859	05141	23
38	91816	08184	93354	06646	94884	05116	22
39	91842	08158	93380	06620	94910	05090	21
40	9.91868	10.08132	9.93406	10.06594	9.94935	10.05065	20
41	91893	08107	93431	06569	94961	05039	19
42	91919	08081	93457	06543	94986	05014	18
43	91945	08055	93482	06518	95012	04988	17
44	91971	08029	93508	06492	95037	04963	16
45	91996	08004	93533	06467	95062	04938	15
46	92022	07978	93559	06441	95088	04912	14
47	92048	07952	93584	06416	95113	04887	13
48	92073	07927	93610	06390	95139	04861	12
49	92099	07901	93636	06364	95164	04836	11
50	9.92125	10.07875	9.93661	10.06339	9.95190	10.04810	10
51	92150	07850	93687	06313	95215	04785	9
52	92176	07824	93712	06288	95240	04760	8
53	92202	07798	93738	06262	95266	04734	7
54	92227	07773	93763	06237	95291	04709	6
55	92253	07747	93789	06211	95317	04683	5
56	92279	07721	93814	06186	95342	04658	4
57	92304	07696	93840	06160	95368	04632	3
58	92330	07670	93865	06135	95393	04607	2
59	92356	07644	93891	06109	95418	04582	1
60	92381	07619	93916	06084	95444	04556	0
'	Cotan	Tan	Cotan	Tan	Cotan	Tan	'
	50°		49°		48°		

242

TABLE 3. LOGARITHMIC TANGENTS AND COTANGENTS

′	42°		43°		44°		′
	Tan	Cotan	Tan	Cotan	Tan	Cotan	
0	9.95444	10.04556	9.96966	10.03034	9.98484	10.01516	60
1	95469	04531	96991	03009	98509	01491	59
2	95495	04505	97016	02984	98534	01466	58
3	95520	04480	97042	02958	98560	01440	57
4	95545	04455	97067	02933	98585	01415	56
5	95571	04429	97092	02908	98610	01390	55
6	95596	04404	97118	02882	98635	01365	54
7	95622	04378	97143	02857	98661	01339	53
8	95647	04353	97168	02832	98686	01314	52
9	95672	04328	97193	02807	98711	01289	51
10	9.95698	10.04302	9.97219	10.02781	9.98737	10.01263	50
11	95723	04277	97244	02756	98762	01238	49
12	95748	04252	97269	02731	98787	01213	48
13	95774	04226	97295	02705	98812	01188	47
14	95799	04201	97320	02680	98838	01162	46
15	95825	04175	97345	02655	98863	01137	45
16	95850	04150	97371	02629	98888	01112	44
17	95875	04125	97396	02604	98913	01087	43
18	95901	04099	97421	02579	98939	01061	42
19	95926	04074	97447	02553	98964	01036	41
20	9.95952	10.04048	9.97472	10.02528	9.98989	10.01011	40
21	95977	04023	97497	02503	99015	00985	39
22	96002	03998	97523	02477	99040	00960	38
23	96028	03972	97548	02452	99065	00935	37
24	96053	03947	97573	02427	99090	00910	36
25	96078	03922	97598	02402	99116	00884	35
26	96104	03896	97624	02376	99141	00859	34
27	96129	03871	97649	02351	99166	00834	33
28	96155	03845	97674	02326	99191	00809	32
29	96180	03820	97700	02300	99217	00783	31
30	9.96205	10.03795	9.97725	10.02275	9.99242	10.00758	30
31	96231	03769	97750	02250	99267	00733	29
32	96256	03744	97776	02224	99293	00707	28
33	96281	03719	97801	02199	99318	00682	27
34	96307	03693	97826	02174	99343	00657	26
35	96332	03668	97851	02149	99368	00632	25
36	96357	03643	97877	02123	99394	00606	24
37	96383	03617	97902	02098	99419	00581	23
38	96408	03592	97927	02073	99444	00556	22
39	96433	03567	97953	02047	99469	00531	21
40	9.96459	10.03541	9.97978	10.02022	9.99495	10.00505	20
41	96484	03516	98003	01997	99520	00480	19
42	96510	03490	98029	01971	99545	00455	18
43	96535	03465	98054	01946	99570	00430	17
44	96560	03440	98079	01921	99596	00404	16
45	96586	03414	98104	01896	99621	00379	15
46	96611	03389	98130	01870	99646	00354	14
47	96636	03364	98155	01845	99672	00328	13
48	96662	03338	98180	01820	99697	00303	12
49	96687	03313	98206	01794	99722	00278	11
50	9.96712	10.03288	9.98231	10.01769	9.99747	10.00253	10
51	96738	03262	98256	01744	99773	00227	9
52	96763	03237	98281	01719	99798	00202	8
53	96788	03212	98307	01693	99823	00177	7
54	96814	03186	98332	01668	99848	00152	6
55	96839	03161	98357	01643	99874	00126	5
56	96864	03136	98383	01617	99899	00101	4
57	96890	03110	98408	01592	99924	00076	3
58	96915	03085	98433	01567	99949	00051	2
59	96940	03060	98458	01542	99975	00025	1
60	96966	03034	98484	01516	10.00000	00000	0
	Cotan	Tan	Cotan	Tan	Cotan	Tan	′
	47°		46°		45°		

243

INDEX

(Numbers refer to pages)